Princples of Biology: Organisms

Biology Lab Manual

First Edition

Jennifer Schramm

Chemeketa Press | Salem, Oregon

Principles of Biology: Organisms
Biology Lab Manual
© 2026 by Jennifer Schramm

ISBN-13: 978-1-955499-46-0

All rights reserved. Edition 1 2026.

No part of this book may be reproduced or transmitted in any form or by any means, electronic or mechanical, including photocopying, recording, or by any information storage and retrieval system, without permission in writing from the publisher.

Chemeketa Press
Chemeketa Community College
4000 Lancaster Dr NE
Salem, Oregon 97305
collegepress@chemeketa.edu
chemeketapress.org

Cover design by Ronald Cox
Interior design by Ronald Cox and Abbey Gaterud

References to website URLs were accurate at the time of writing. Neither the author nor Chemeketa Press is responsible for URLs that have changed or expired since the manuscript was prepared.

Printed in the United States of America.

Land Acknowledgment
Chemeketa Press is located on the land of the Kalapuya, who today are represented by the Confederated Tribes of the Grand Ronde and the Confederated Tribes of the Siletz Indians, whose relationship with this land continues to this day. We offer gratitude for the land itself, for those who have stewarded it for generations, and for the opportunity to study, learn, work, and be in community on this land. We acknowledge that our College's history, like many others, is fundamentally tied to the first colonial developments in the Willamette Valley in Oregon. Finally, we respectfully acknowledge and honor past, present, and future Indigenous students of Chemeketa Community College.

Contents

Lab and Safety Regulations

Personal Behaviors

1. Eating and drinking are prohibited in the labs.
2. Wash your hands before leaving the lab.
3. Wash the lab counter & related work areas with disinfectant before and after the lab.
4. Stow your personal items in the cubbies or under your station.
5. Keep the lab counters and work areas uncluttered.
6. Clean-up is your responsibility:
    a. Return cleaned items back to lab kits.
    b. Wash slides and return to the original location (unless told otherwise). Cover slips may be thrown away.
    c. Wash glassware and return to original location (unless told otherwise).
    d. Ensure any lab waste is disposed of in the approved container(s).
    e. Ensure sinks and counters are as clean as when you arrived.
7. Clothing must be appropriate for the lab to be performed.
8. Read the lab in advance so you're aware of proper safety protocols.
9. If you have questions about safety, ask before you do.

Safety Protocols

1. Know where the lab safety equipment is located–fire extinguisher, first aid, eye wash, etc.
2. Wear personal safety equipment (goggles, gloves, aprons) as indicated in the lab instructions or by your instructor.
3. Handle all chemicals and biologicals, including stains, below eye level.
4. Do not assume something can just go down the sink or in the trash. Dispose of wastes in appropriate waste containers.
5. If there is a chemical spill:
    a. Get your instructor.
    b. If the spill is on the floor or counter, keep away from it.
    c. If the spill is on you or your clothes, rinse the area immediately with running water.
    d. If you splash chemicals in your eyes, go immediately to the eye wash station, turn on cold water, remove the red caps and lean down so that the water bubbles into your eyes.
6. In case of injury:
    a. Get your instructor.
    b. If the instructor is not available, call campus safety (x5023) on the lab's phone.
7. If the injury appears severe, also call 911
8. In case of fire:
    a. If your clothes are on fire, yell "FIRE" and roll on the floor or use a coat or fire blanket to smother it.
    b. If chemicals or lab equipment is on fire, call out "FIRE" and evacuate the room.
9. If the fire alarm goes off, take your essential items and leave the room. Head to the evacuation area as told by your instructor. Stay with your class and wait for instructions. Do not re-enter the building until an all-clear is given.

1 | Comparative Anatomy Pre-Lab

Instructions

- ❑ Read the lab manual and then follow the instructions to complete the assignment.
- ❑ You may need your textbook and other resources to complete this assignment.
- ❑ Pre-labs must be completed before the start of the lab.

1. Explain the phrase "structure equals function" and provide an example to support your explanation.

2. Why do animals have tissues while plants have tissue systems?

3. Complete Table 1.1 to differentiate between the three tissue systems found in plants.

Table 1.1. Tissue Locations and Functions in a Plant Tissue System

Plant Tissue System	Location in the plant	Function
Dermal		
Ground		
Vascular		

4. Complete Table 1.2 to compare the four types of animal tissue.

Table 1.2. Animal Tissue Types

Tissue	Basic structure	Function
Connective		
Epithelial		
Muscle		
Nervous		

5. How many types of tissue/tissue systems are found in each of the following?

___ Leaf

___ Heart

___ Skin

___ Root

1 Comparative Anatomy: Tissues & Organs

Plants and animals are both multicellular organisms. Multicellular organisms contain many different types of cells that perform unique functions in the body. Those cells are further organized into **tissues**, where they work with other cells to perform specific tasks. Groups of tissues are arranged into **organs** that are structures that perform specific functions in key organ systems in the body. **Organ systems** are groups of organs arranged to perform specific tasks that are key to maintaining homeostasis.

By the end of this lab, you should be able to:

- ❑ Distinguish between the way plant and animal tissues are organized.
- ❑ Determine whether a specimen at cell-level magnification is from a plant or animal.
- ❑ Identify the key plant organs and relate them to the structure of animal organs

Exercise 1.1. Animal Tissues

Animals have four basic tissue types. As you examine these slides, remember that even though the slides are labeled as a tissue type, they are actually sections through organs. Because of this, there are generally several different types of tissues present on a slide.

> **If you are having trouble identifying the cells/tissues** that you are supposed to be seeing, please ask – it doesn't pay to be observing a slide for 10 minutes only to find out that you are looking at the wrong thing.

Figure 1.1. Animal Tissue Organization. Terminology mentioned in the text is labeled in each diagram.

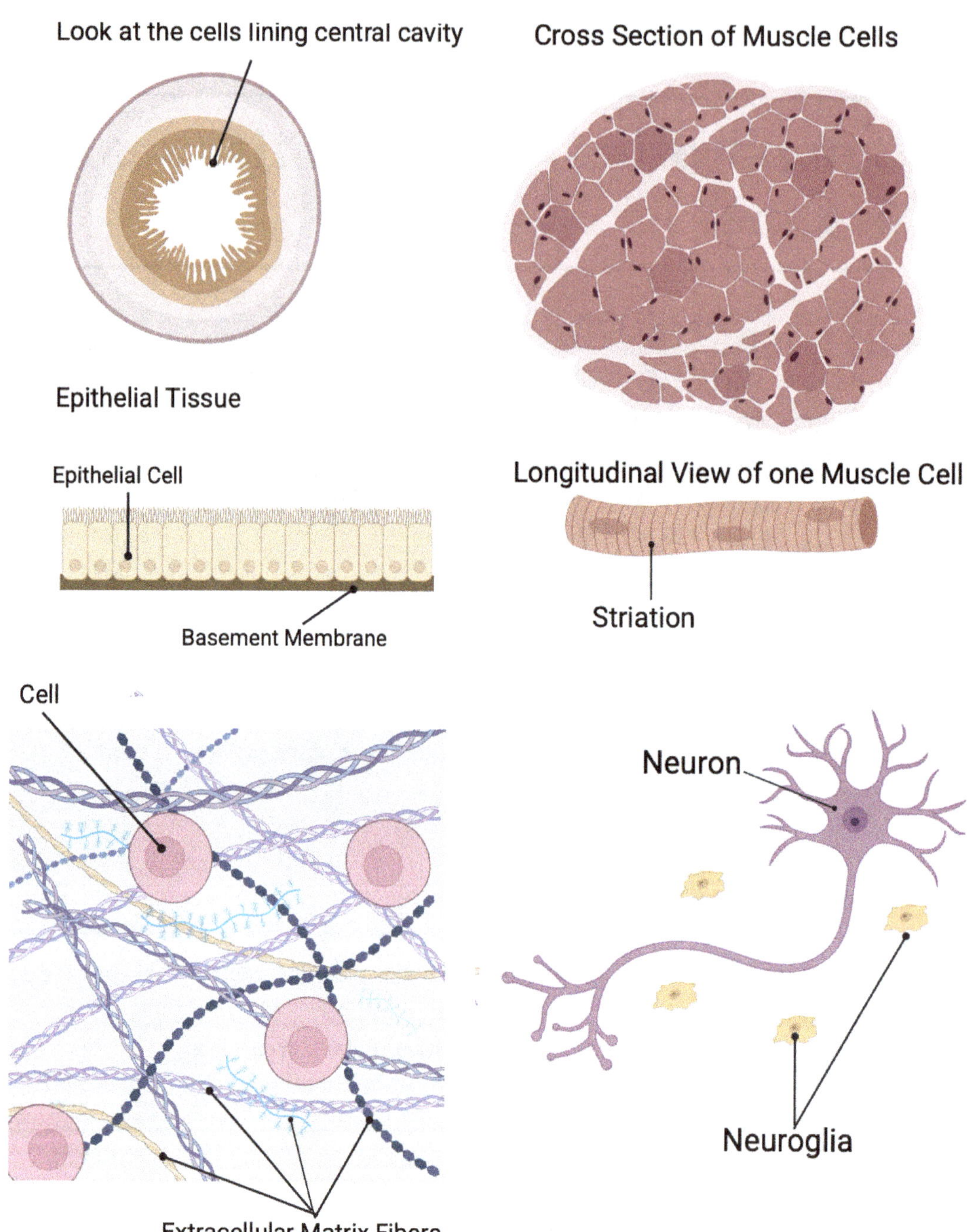

A. Observing Animal Tissues

Instructions

1. Work with a partner to examine each of the prepared slides listed below.
2. Create a scaled sketch of 5-10 cells from each slide and record your observations in your lab notebook.
3. Label any organelles or structures you can identify.

A. Epithelial Tissue

Key traits (See Figure 1A):

- ❑ One surface is exposed to an open space, and the other surface is attached to a thin basement membrane.
- ❑ Cells are tightly attached to each other.
- ❑ Epithelial types are classified by the number of cell layers and shape.

Slide name: Simple Columnar Epithelium

- ❑ This slide usually contains a cross-section of the intestine. It will look like a donut. Look at the cells lining the "donut hole."
- ❑ Epithelial cells line surfaces (they will not be embedded in the middle of a bunch of cells).

B. Muscle tissue

Key traits (See Figure 1B):

- ❑ Cells are usually very long and often appear stringy.
- ❑ In some types of muscle, you may see bands of dark and light known as **striations**.

Slide name: Muscle Types

- ❑ The slide includes 3 types of muscle tissue. Do not be concerned about trying to tell them apart. Focus on the key features of all muscles!
- ❑ At high magnification, in the right type of muscle, you may be able to see **striations.**

C. Connective Tissue

Key traits (See Figure 1C):

- ❑ Generally consists of fewer cells surrounded by a large amount of extracellular matrix (ECM).
- ❑ The ECM contains protein fibers (threads) embedded in a solid, gel, or liquid.

Slide name: Areolar Connective Tissue

- ❑ View this slide with your high-power objective.
- ❑ Cells and fibers should be visible.

D. Nervous tissue

Key traits (See Figure 1D):

- ❑ Includes large branching cells (**neurons**) and small supporting **neuroglia**

Slide name: Neuron Smear

- ❑ Not an intact tissue; cells have been spread out to make the structures clearer.
- ❑ Large purple cells with long extensions that trail off are neurons.
- ❑ Blue dots around the neuron are the nuclei of neuroglial cells that help to care for the neuron

Exercise 1.2. Animal Organs: The Skin

In animals, organs are formed from at least two types of animal tissue. These tissues work together to perform a specific job in the body, usually as part of an organ system. A great example of an organ is the skin, the largest and most prominent organ in the mammalian body. The skin contains all four types of animal tissue.

Slide name: Mammalian Skin (scalp) & Skin Model
- ❏ The top (facing the environment) has hairs sticking out
- ❏ Hairs are supported by tiny muscles embedded in connective tissue.
- ❏ Nervous tissue forms onion-like pads in the connective tissue

A. Examining Mammalian Skin

Instructions
1. Examine the slide of mammalian skin.
2. Use the skin model, its key, and your observations to identify the locations of each type of animal tissue (connective, epithelial, muscle, and nervous) in the skin.
3. Use your observations to identify and label the four animal tissue types (connective, epithelial, muscle, and nervous) in the diagram provided in Figure 1.2.

Figure 1.2. Diagram of a section of skin.

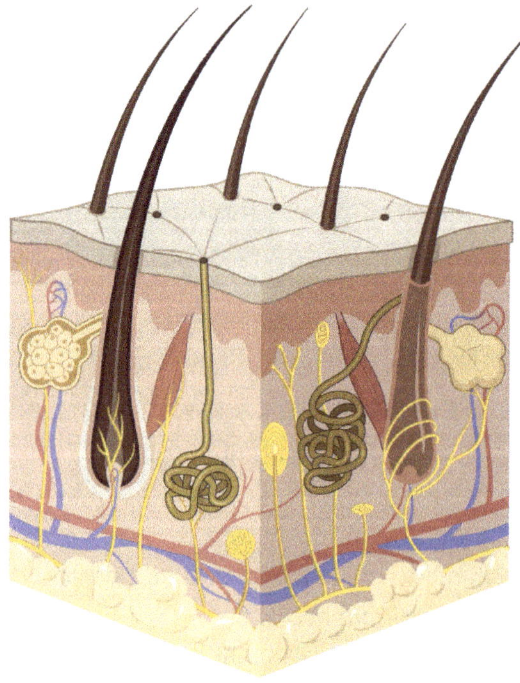

Q1. Identify and label the four animal tissues in Figure 1.2.

Q2. What features can you use to distinguish between a plant cell and an animal cell? (you may want to revisit your sketch of a check cell from our first lab this term.)

Q3. What tips would you give to someone who is trying to tell the four types of animal tissues apart?

Q4. Use what you learned about tissues and skin to complete Table 1.3.

Table 1.3. Skin Functions and Tissue Type

Skin function	Tissue Types involved	How the skin structure facilitates this function
Protection against microbe infection		
Prevention of dehydration		
Regulation of body temperature		
Reception of stimuli from the surrounding environment		

Exercise 1.3. Plant Organs and Tissue Systems

Animals have tissues, and plants have **tissue systems**. A tissue system is a group of cells with a specific structure and organization that extends through the body of the plant from root to shoot. Plants contain three tissue systems: **dermal, vascular, and ground.** The **dermal tissue system** is represented by the plant's surface and usually consists of a single layer of tightly packed cells that protect the plant from the environment. The **vascular tissue system** forms a series of continuous tubes that move water, minerals, and carbohydrates around the plant. The **ground tissue system** fills the space between the other two tissue systems and exhibits a variety of functions, including storage, support of plant structure, and photosynthesis.

The best way to explore plant tissue systems is to examine the plant's main organs. In this activity, we will examine the arrangement of these tissue systems in the plant stem and then apply what we have learned to the leaf and the root.

A. Examining a Plant's Major Organs

Instructions

1. Work with a partner to examine each of the prepared slides listed below
2. Create a scaled sketch (overview rather than every cell; see cross sections in Figure 1.2 for an example) of each specimen.
3. Label the three tissue systems.

> Note on these slides: Most of them show two different specimens. Use the provided description to determine which specimen you should examine, but posters hanging on the classroom wall can also aid in orientation and identification.

Figure 1.3. Layout of tissues in plant organs. A= Stem, B=Root, and C=leaf.

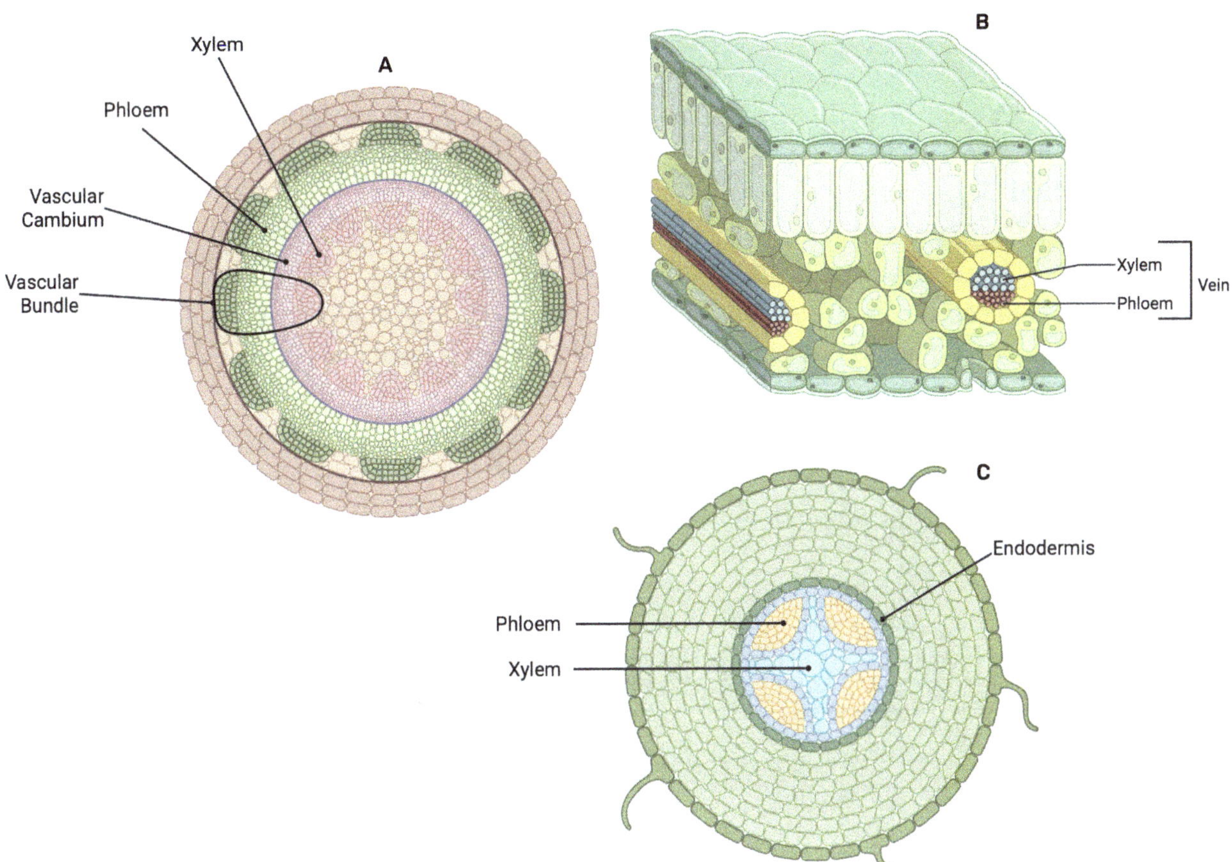

Plant Organ: Stems

Key Traits (See Figure 1.3A):

- ❑ Vascular tissues are arranged into oval to triangular striped vascular bundles.
- ❑ Xylem and phloem are separated by a few rows of "blurry" cells called the vascular cambium.
- ❑ Vascular bundles are arranged into a ring just below the dermal tissue.

Slide name: Monocot and Dicot Stems

- ❑ Use the 4X objective to identify the three tissue systems in the specimen.
- ❑ Use Figure 1.1A and other resources provided by your instructor as guidance.

Plant Organ: Roots

Key Traits (Figure 1.3B):

- ❑ Vascular tissue is arranged into a centralized bundle surrounded by thick-walled cells (endodermis = "inner skin").
- ❑ Large vascular cells within the core form a cross or star pattern.

Slide name: Monocot and Dicot Roots

- ❑ Use the 4X objective to identify the three tissue systems in the specimen.
- ❑ Use Figure 1.1B and other resources provided by your instructor as guidance.

C. Plant Organ: Leaf

Key Traits (Figure 1.3C):

- ❑ Tissue arranged into a sandwich
- ❑ Vascular tissue forms a central "vein" throughout the sandwich
- ❑ Two distinct layers of ground tissue (columnar cells and irregularly shaped cells)

Slide name: Syringa Leaf Cross-section

- ❑ Use the 10X objective to identify the three tissue systems in the specimen.
- ❑ Use Figure 1.3C and other resources provided by your instructor as guidance.

Exercise 1.4. The Plant Body

The body of a plant is divided into the **shoot**, the above-ground region, and the **root**, the below-ground region. The primary role of the shoot portion of a plant is to perform photosynthesis. The structure of the root, on the other hand, has evolved to ensure the efficient uptake of water and minerals from the soil. This activity takes a closer look at the structure of a plant.

A. Examining Plant Structures

Instructions

1. **Examine a carrot.** The intersection of green and orange is where the shoot and root meet in this plant. Carrots are bred to have a large and colorful tap root, but you may see secondary roots (root hairs are only visible with a microscope).
2. **Examine a stem of asparagus.** The small triangles along the stem are leaves. Use these features to identify a node, an internode, and a lateral bud.
3. **Examine the seedling** provided by your instructor. Identify the root and shoot, a node, an internode, the tap root, and a secondary root.
4. **Label these features** in Figure 1.3.

Q5. What do lateral buds develop into? How is the ability to develop this feature important to the success of the plant?

Q6. Where are the internodes in the carrot?

Q7. Why do plants have secondary roots?

Exercise 1.5. Connecting Roots to Shoots

A. Observing the Interconnectedness of Plant Tissue Systems

A key feature of plant tissue systems is their interconnectedness throughout the plant's body. Use your observations of these three slides to complete Figure 1.4.

Instructions
1. **Label the following features** of the seedling on the left side of Figure 1.4: internode, lateral bud, node, root, and secondary root. shoot, tap root.
2. Choose three distinct colored pencils.
3. **Complete the key in Figure 1.4** to associate a plant tissue system with a specific color.
4. Color the diagrams (Figure 1.4) on the right to indicate which parts of the organ represent which tissue system.

Figure 1.4. Organization of the plant body.

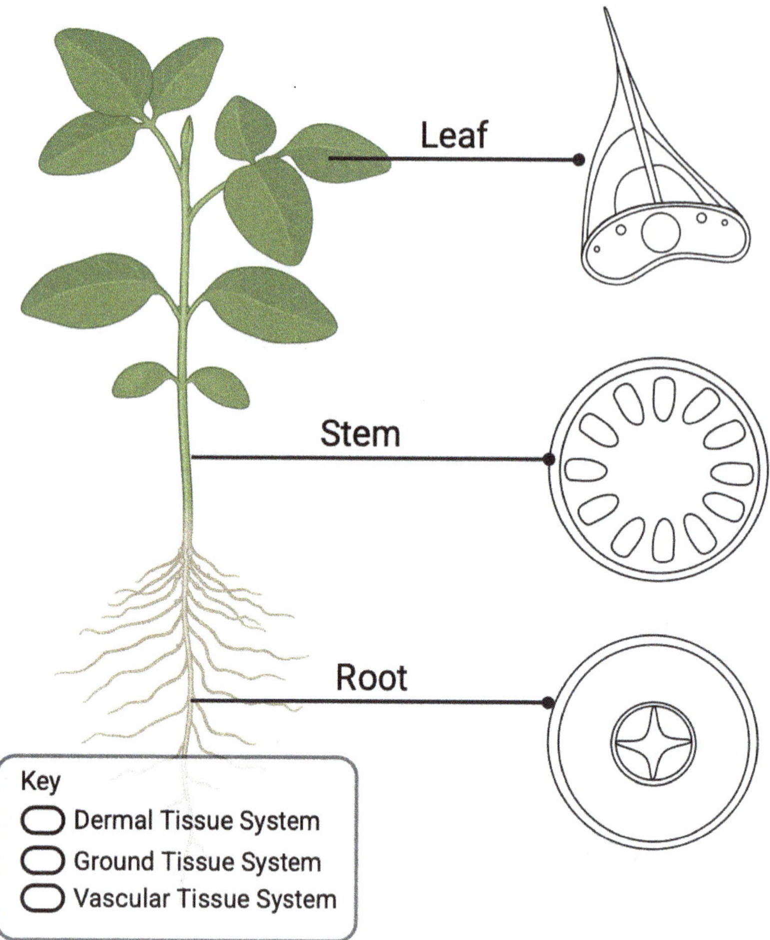

Q8. What features can you use to determine if the tissue on a slide is from a plant or an animal?

Q9. Are any plant tissues analogous to animal tissues?

Q10. Complete Table 1.4 to relate the structure of the leaf to its function in the plant.

Leaf function	Tissue Types involved	How leaf structure facilitates this function
Capture and transform sunlight		
Prevention of dehydration		
Gas Exchange		
Protect against invaders		

Exercise 1.6. Grocery Store Botany

Fruits and vegetables are a big part of our lives, but botanists have a different definition of fruits and vegetables than a chef might. In cooking, we classify plants we consume as fruit or vegetables primarily by taste, but botanists define **fruit** as any plant product that develops from the ovary of a flower. Fruits always contain seeds. According to a botanist, **vegetables** are plant foods derived from other plant organs, including stems, leaves, roots, buds, and even immature flowers.

A. Identifying Fruits or Vegetables

In this exercise, we will use what we have learned about the structure of the various plant organs to identify plant foods as fruit or vegetables.

Instructions
1. Examine each plant food provided in the lab.
2. For each plant part, determine whether the food is a fruit or a vegetable based on the botanical definition.
3. If it is a vegetable, determine what plant organ it represents (stem, root, leaf, bud, immature flower, etc.).
4. Complete Table 1.5 based on your observations.

Table 1.5. Plant Systems and Vegetables

Plant Food	Fruit or vegetable?	VEGGIES ONLY: What plant organ does it represent?	How do you know? (provide at least 3 bullet points)
Broccoli			
Celery			
Onion			
Eggplant			
Asparagus			
Potato			
Sweet potato			

1 | Applying What You've Learned

A selection of these questions will appear on your lab quiz. Lab quizzes are open-book. Type up answers to these questions in advance so that you can copy and paste the answers into the quiz.

1. How does the arrangement of tissues in the skin contribute to the overall function of this organ?

2. What is the primary difference between the tissues of animals and the tissue systems of plants?

3. Choose one tissue or tissue system and relate its structure to the function it performs.

2 | Homeostasis Pre-Lab

Instructions

- ❏ Read the lab manual and then follow the instructions to complete the assignment.
- ❏ You may need your textbook and other resources to complete this assignment.
- ❏ Pre-labs must be completed before the start of the lab.

1. Describe the role of the circulatory system in maintaining homeostasis.

2. How might a change in heart rate impact homeostasis?

3. In this lab, you will have the opportunity to observe the impact of a specific substance on homeostasis in blackworms. Refer to your textbook and/or look up the following substances to complete Table 2.1 and learn more about your options.

Note: Not all substances may be available for your lab.

Table 2.1. Substance Effect on Homeostasis

Substance	Biological Role/Impact	Predicted impact on heart rate (+, - or no impact)
Acetylcholine		
Pseudoephedrine or Phenylephrine		
Lidocaine		
pH		
Mg^{2+} ions		
Caffeine		
Alcohol		
Salt		
Nitrate		

2 **Homeostasis in Blackworms**

One of the key features of life is the ability to maintain homeostasis. The cells, organs, and tissues in the body of a multicellular organism alter their functions to ensure that temperature, osmolarity, pH, and substrate concentrations are at the optimal level for enzyme function and, ultimately, for organism survival.

To maintain the optimal level, organisms have set points and mechanisms that return the body to this state when conditions are out of balance. These systems function much like the thermostat that heats your home.

You set the temperature of your home (set point), and the thermostat monitors it. When the room is colder than your set point, the thermostat turns on the furnace. When the temperature goes above your set point, the thermostat turns the furnace off.

The body is much more complex than a thermostat and contains thousands of different set points and mechanisms that work to maintain optimal conditions within the organism. Unlike a thermostat, the mechanisms that maintain homeostasis in an organism do not function independently. Instead, the nervous and endocrine system function to coordinate the actions of all body systems in order to maintain optimal conditions.

By the end of this lab, you should be able to:

- ❏ Connect changes in heart rate to maintenance of homeostasis.
- ❏ Describe the role of the nervous and endocrine systems in maintaining homeostasis.
- ❏ Indicate several environmental factors that impact homeostasis.

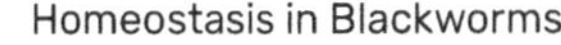

Exercise 2.1. Measuring Blackworm Heart Rate

A. Using Blackworm Hearts to Study Homeostasis

In this lab, we will use blackworm heart rates to study homeostasis. The heart plays a key role in homeostasis, and its rate of pumping determines how quickly resources and signals are delivered to the body's tissues.

The heart has a **resting rate** which is equivalent to the set point of the circulatory system. Environmental stimuli that move systems away from their set points can raise or lower the heart rate to help other systems in the body return to their set point more efficiently.

Blackworms are transparent annelids (segmented worms). With the aid of a light microscope, one can easily visualize the pulsing muscular dorsal aorta (blood vessel) that runs the length of the worm. The aorta acts as a heart, pumping blood and nutrients into the body cavities through a pair of arteries found in each segment. Blackworm heart rate can easily be measured by counting the pulses of the aorta in one segment of the worm.

Instructions
1. Fill the plastic beaker with conditioned water.
2. Rinse and dry 3 regular slides and 3 depression slides with conditioned water.
3. Label 5 small petri dishes with the letters A – E and fill each with conditioned water.
4. Label a large petri dish "stock" and fill with water from the stock tank.
5. Take the large petri dish to the blackworm stock tank. Use a pipette with the tip cut off to obtain 5 black worms of approximately the same size. Place the worms in the large petri dish and return to your work area.
6. Working quickly
   a. Use a pipette to transfer one worm and a small amount of water to a clean flat slide.
   b. Use the edge of the coverslip to cut the ends off the worm (about ¼ of the worm should be removed from each end).
   c. Return the ends of the worm to the stock dish.
   d. Use a pipette filled with conditioned water to rinse the center segment of the worm into one of the 5 small petri dishes labeled A – E.
   e. Place a drop of conditioned water into the well of a depression slide. Use a pipette to transfer the worm segment from dish A to the well of the depression slide.
   f. Cover the well with a coverslip.
7. Examine the slide at low power on your compound microscope. Identify the dorsal aorta and look for the pulsations.

Determine the resting heart rate

8. Center one section of the aorta in the middle of your field of view.
9. While a partner keeps time, count the number of pulses that move through that segment in 20 seconds.
    a. Record your result in Table 2.1 (at the end of the handout; worm A, measurement 1).
    b. Wait 30 seconds and recount.
    c. Record your results in Table 2.1 (worm A, measurement 2).
    d. Wait 1 more minute and recount.
    e. Record your results in Table 2.1 (worm A, measurement 3).
10. Repeat for worms B and C.
11. Rinse your worms and the contents of the stock dish into the recovery tank.
12. Complete Table 2.1 by adding the three 20-second measurements for each worm together to obtain beats per minute.

> Note: Worms D and E are backups to use if a segment dies during the experiment.

Table 2.1. Blackworm Heart Rate

	Number of pulses in 20 seconds			Normal Heart Rate (BPM)
Worm Sample	Measurement 1	Measurement 2	Measurement 3	
A				
B				
C				

Q11. Why do we add the measurements together rather than averaging them?

Q12. Did your three worms have the same normal heart rate? Why or why not?

Q13. Would it be appropriate to average the normal heartbeat of the three worms? Why or why not?

Exercise 2.2. Changing the Blackworm Heart Rate

In this exercise, you will design an experiment that will alter the heart rate of a blackworm.

A. Designing an Experiment

The following options may be available (check with your instructor). No group should replicate the experiment of another group

- ❑ If you do not use the same worms used in Exercise 2.1, plan to record the resting heart rate for each worm before exposing it to the treatment.
- ❑ Substances can be added to the worm by removing the water in the depression slide with a pipette and replacing it with the solution you are testing.
- ❑ Allow the worms to soak in the testing solution for 5 minutes before taking. measurements.

Table 2.2. Available Treatments

0.1 mg/ml Acetylcholine	5 mM caffeine
~1 mM Pseudoephedrine or Phenylephrine	0.02% ethanol
1 mM lidocaine	0.6% NaCl (recommend dilution)
10 mM sodium phosphate at pH 5 or pH 8	1 mg/L NaNO$_3$
1 mg/L MgCl$_2$	

Instructions

On a dry-erase board
1. Indicate the treatment you would like to study.
2. Provide a brief explanation of how that treatment relates to homeostasis.

Obtain your instructor's approval before moving forward.

On paper

3. Define the question you would like to ask.
4. State your hypothesis and the prediction for your experiment in your lab notebook.

On a dry-erase board

5. Write out the step-by-step procedure you plan to perform.

Hint: Use the setup for Exercise 2.1 as a template and modify it according to the instructions provided at the start of Exercise 2.2.

Obtain your instructor's approval before moving forward.

6. Run your experiment and record your data in Table 2.3.

Table 2.3. Effect of _______________________________________ on Blackworm Heart Rate

Worm	Measurement #	Number of Pulses in 20 sec	Number of pulses in 20 seconds after Treatment with ________________
A	1		
A	2		
A	3		
B	1		
B	2		
B	3		
C	1		
C	2		
C	3		

Exercise 2.3. Analyzing Your Results

Individuals of the same species have similar, but not identical characteristics. One factor that can vary among individuals is heart rate. This difference makes it challenging to compare the raw data you collected in your experiment. To more accurately assess the results in this situation, the data must be normalized by calculating the percentage change relative to the normal heart rate.

A. Calculating the Percentage Change

Instructions

1. Complete Table 2.4.
    a. Add the three 20-second measurements for each worm in Table 2.4 together to obtain beats per minute.
    b. Determine the percent change in heart rate using the following formula:

    [(Rate with drug – Normal Rate)/Normal Rate] x 100%

2. Determine the average % change by subtracting the normal heart rate from the heart rate after treatment, dividing by the normal heart rate and multiplying by 100.
3. Summarize your results and conclusions in your lab notebook.

> Note: A negative result means that the heart rate dropped. A positive result indicates that the heart rate increased.

Table 2.4. Normalized Change in Heart Rate After Treatment with _________________________

Worm	Normal Heart Rate (BPM)	Heart Rate After Treatment (BPM)	% Change in Heart Rate After Adding __________
A			
B			
C			
Average change in Heart Rate:			

Q14. How is the treatment you planned for Exercise 2.2 related to homeostasis in the blackworm?

Q15. How did your experimental treatment affect the heart rate of blackworms?

Q16. Does your data support that blackworm homeostasis is affected by your treatment?

2 | Applying What You've Learned

A selection of these questions will appear on your lab quiz. Lab quizzes are open-book. Type up answers to these questions in advance so that you can copy and paste the answers into the quiz.

1. Homeostasis is about maintaining constant internal conditions. If that's the case, how can changing the heart rate maintain homeostasis?

2. If you used different worms in Exercises 2.1 and 2.2, you were instructed to determine the normal pulse rate of each worm before performing the experimental treatment in Exercise 2.2. Why would it be inappropriate to use the normal pulse rates in Table 1 (results of Exercise 2.1) as your baseline for comparison?

3. Predict what would happen to the pulse rate of a blackworm in each of the following situations. For each prediction, provide a scientific explanation of your reasoning.

   ❑ The worm is placed in an anaerobic environment.
   ❑ The worm is incubated at different temperatures.
   ❑ The worm is incubated in a dark room.

<table>
<tr><td>3</td><td><h1>Fermentation Pre-Lab</h1></td></tr>
</table>

Instructions

- ❑ Read the lab manual and then follow the instructions to complete the assignment.
- ❑ You may need your textbook and other resources to complete this assignment.
- ❑ Pre-labs must be completed before the start of the lab.

1. Complete Table 3.1 comparing the stages of cellular respiration.

Table 3.1. Comparing Glycolysis and Aerobic Cellular Respiration

	Location in cell	Aerobic or Anaerobic?	Substrate	Product
Glycolysis				
Aerobic Cellular Respiration				

2. Which of the processes listed above are necessary for fermentation?

3. Complete Table 3.2 to compare cellular respiration to fermentation.

Table 3.2. Comparing Cellular Respiration and Fermentation

	Location in cell	Aerobic Cellular Respiration	Substrate	Product
Cellular Respiration				
Fermentation				

4. Which process (cellular respiration or fermentation) produces more ATP? Why bother doing the process that doesn't produce as much ATP at all?

5. To determine the rate of a chemical reaction, we can either determine substrate disappearance or product appearance. Which of these factors are we measuring in this lab activity?

6. Use your textbook and the Internet to gather information about the best environment for Baker's yeast (*Saccharomyces cerevisiae*; literally "sugar-eating fungus"). Suggestions provided below, but the more diverse the information we gather, the better our inquiry experiments will be.
   - ❑ What is the optimal temperature for this species?
   - ❑ What is the optimal pH of this species?
   - ❑ What other information can you learn about this species that might help you design an experiment?

   Provide at least 3 facts. ____

3 Alcoholic Fermentation

One of the most important catabolic reactions in cells is the breakdown of glucose to produce ATP. The breakdown of glucose can take place in two ways, depending on whether oxygen gas (O_2) is present. **Glycolysis**, the first metabolic process to evolve for this purpose, occurs in all cells in the absence of O_2 and produces a net of 2 ATP per glucose catabolized. In cells that only perform glycolysis, fermentation must also occur.

Glycolysis relies on the energy carrying molecule NADH. In its reduced form, NADH holds excited electrons and is high in energy. In its oxidized form, NAD+ does not carry electrons and is lower in energy. During glycolysis the cell draws on a limited pool of NAD+, converting it to NADH. Without additional NAD+, the cell cannot continue to break down sugar. The process of fermentation converts NADH back to NAD^+, thereby allowing glycolysis to continue.

By the end of this lab, you should be able to:
- ❑   State the chemical reaction for alcoholic fermentation and indicate the substrates and products.
- ❑   Explain how a fermentation tube can be used to study the process of cellular respiration.
- ❑   Design an experiment to alter the rate of fermentation in yeast.

Exercise 3.1. Fermentation

Yeast is well known as an organism that undergoes alcoholic fermentation. The three carbon sugar, pyruvate, produced by glycolysis is converted to alcohol. In the process, pyruvate is reduced with electrons from NADH, recycling NAD+ for glycolysis, and carbon dioxide (CO_2) is produced. Humans take advantage of alcoholic fermentation in yeast to make alcohol (beer, wine, etc.) and bread.

A. Examining the Process of Alcoholic Fermentation

In this exercise, we will examine the process of alcoholic fermentation by altering the amount of yeast we put in a special "fermentation tube" (Figure 3.1).

In a fermentation tube, a reaction mixture is added to the "bowl" of the tube, and then the tube is tilted, forcing air bubbles out of the "tail" until it is full.

When the tube is upright, the solution in the tail is no longer exposed to O_2, and the organisms in this region of the tube undergo fermentation. As the yeast produces CO_2, it rises to the tip of the "tail" of the fermentation tube, lowering the solution level. Over the course of the activity, we will measure the rate of fermentation by observing how quickly the tail fills with gas.

Instructions

1. Number the fermentation tubes 1-4 using a wax pencil.
2. Add solutions as indicated below in Table 3.3. DO NOT invert the tubes until all four tubes are ready.

Table 3.3. Numbered Tubes and Solutions

Tube #	DI Water (ml)	Yeast Solution (ml)	Glucose Solution (ml)
1	12	0	9
2	18	3	0
3	9	3	9
4	3	9	9

Q1. Which of the treatments in this experiment are experimental and which are controls?

Q2. Does this experiment include positive and negative controls? How do you know?

Q3. Why does each tube contain a different amount of water?

3. Place your thumb or a piece of Parafilm over the opening to each fermentation tube. Slowly invert the tube until the tail is full of solution.
4. Stand the tube upright.
5. Mark the top of the water column with a wax pencil. If everything has worked well, the water column should extend to the top of the tail. If the water column is more than ¼ of the tail's length away from the top, try again.
6. Place the fermentation tubes into a 30°C water bath. Make sure the opening of the fermentation tube is above the water level of the water bath.
7. Record the mm of gas accumulating in the tail at 2-minute intervals for 20 minutes in Table 4.4. *Measure from the meniscus, which may be below the bubbles.*
    a. Determine the amount of CO_2 by measuring the mm of space filled with CO_2. If the water column extends all the way to the tip of the tube at the start of the experiment, measure from there to the meniscus.
    b. If you started below the end of the tube and marked the starting point, measure from the starting point to the meniscus.

Table 3.4. Effect of Yeast and Glucose Concentration on the Rate of Fermentation (Exercise 3.1)

Time (min)	Amount of CO_2 Accumulated in Tail of Fermentation Tube (mm)			
	Tube #1 (no yeast)	Tube #2 (no glucose)	Tube #3 (3 ml yeast solution)	Tube #4 (9 ml yeast solution)
2				
4				
6				
8				
10				

12				
14				
16				
18				
20				

8. Graph your results on a piece of graph paper.

Q4. What conclusion can you draw from your data regarding how changing yeast concentration affects the rate of fermentation?

Q5. Does your data support your hypothesis? Why or why not?

Exercise 3.2. Changing the Rate of Fermentation

In this activity, you will design an experiment that will alter the rate of fermentation using the same procedure you used in Exercise 3.1.

Your instructor will provide a variety of solutions and environments that you might test. Additional options may be available, so be creative.

Note: Every group must perform a different experiment.

- ❏ What question is your group asking?
- ❏ What is your hypothesis? *Make sure it answers the question you posed.*
- ❏ What do you think will happen in your experiment if your hypothesis is valid?

On a dry-erase board
1. Write out the step-by-step procedure you plan to perform.

Hint: Use the set-up for Exercise 3.1 as a template. That experiment changed the amount of yeast. This time, use the same amount of yeast in every sample and change just one other factor.

2. Complete Table 3.5 by finishing the title and adding treatment names that reflect your experiment.
3. Run your experiment.
    a. Follow the procedure you outlined on the dry-erase board.
    b. Record your data in Table 3.5.

Table 3.5. Effect of_________________________on the Rate of Fermentation (Exercise 3.2)

	Amount of CO_2 Accumulated in Tail of Fermentation Tube (mm)			
Time (min)	Tube #1	Tube #2	Tube #3	Tube #4
2				
4				
6				
8				
10				
12				
14				
16				
18				
20				

4. Graph your results on a piece of graph paper.
    a. Each student should create their own graph.
    b. Graphs produced electronically are permitted if they are imported into a word processing program so a figure legend can be added.
    c. Figures should include a complete figure legend.
    d. Follow all the graphing requirements laid out in the Scientific Writing Handout.
5. Prepare a scan of the data table and graph for this experiment for the lab quiz.

Q6. What conclusion can you draw from your data regarding how the independent variable affects the rate of fermentation?

3

Q7. Does this data support your hypothesis? Why or why not?

3 | Applying What You've Learned

A selection of these questions will appear on your lab quiz. Lab quizzes are open-book. Type up answers to these questions in advance so that you can copy and paste the answers into the quiz.

1. When making bread, yeast is mixed with lukewarm water and, sometimes, a bit of sugar. Yeast is mixed into the flour to form dough, which is then allowed to rise. Use what you learned about fermentation in this lab to explain why bread rises.

2. Yeast is also used in beer-making. In that process, yeast is mixed with ground-up barley (or other grains). The grain is a rich source of sugar. Explain how yeast converts this mix of water and grain to a bubbly alcoholic beverage.

3. In lecture, we learned about cellular respiration, but this lab explores the process of fermentation. What is the connection? Differentiate between these two processes.

<table>
<tr><td>**4**</td><td># Photosynthesis & Cellular Respiration Pre-Lab</td></tr>
</table>

Instructions

- ❑ Read the lab manual and then follow the instructions to complete the assignment.
- ❑ You may need your textbook and other resources to complete this assignment.
- ❑ Pre-labs must be completed before the start of the lab.

Algae Beads

Download the Spectral Analysis Application onto a device you bring to the lab (at least one person in your group will need to have the app). Use the link provided on the course website to download the app.

Watch the video provided in QR Code 4.1 before answering the following questions.

QR Code 4.1. Experimental Setup for the Algae Bead Lab to Study Photosynthesis and Cellular Respiration.

1. Why does the solution of the CO_2 indicator change color during the photosynthesis and cellular respiration of algae beads?

2. Enzyme-catalyzed reactions require an enzyme, substrate(s), and product(s). Complete Table 4.1 by indicating which ingredient represents which part of the algae bead reaction we are performing in this lab?

Table 4.1. Ingredients and Algae Beads

In this experiment, the…..	….is represented by the….
Enzyme	
Substrate	
Product	

3. The rate of a chemical reaction is determined by measuring substrate disappearance or product appearance. What are we measuring in this reaction?

4. Which process (photosynthesis, cellular respiration, or both) do the algae perform when incubated in the light? In the dark?

5. What should happen to the pH of a solution as the algae perform photosynthesis?

6. What should happen to the pH of a solution as the algae perform cellular respiration?

Chromatography

7. What is the role of pigments in the process of photosynthesis?

8. Describe the process of paper chromatography.

9. Figure 5.2 in this lab suggests that you count the number of "polar oxygen atoms" in each pigment molecule. What is a polar oxygen atom?

Name: _________________________________ Lab Time: _______________ Due: __________________

4 Photosynthesis & Cellular Respiration

By the end of this lab, you should be able to:

- ❏ Explain how chromatography is used to separate plant pigments by polarity.
- ❏ Use a spectrophotometer to measure changes in absorption.
- ❏ Explain how a pH indicator can be used to measure the rates of photosynthesis and cellular respiration.
- ❏ Describe several factors that affect the rate of photosynthesis.

Exercise 4.1. Algae Beads

In this experiment, we will examine the rates of photosynthesis and cellular respiration in *Scenedesmus obliquus* cells (Figure 4.1) using a CO_2 indicator.

Figure 4.1. *Scenedesmus obliquus* **cells.**

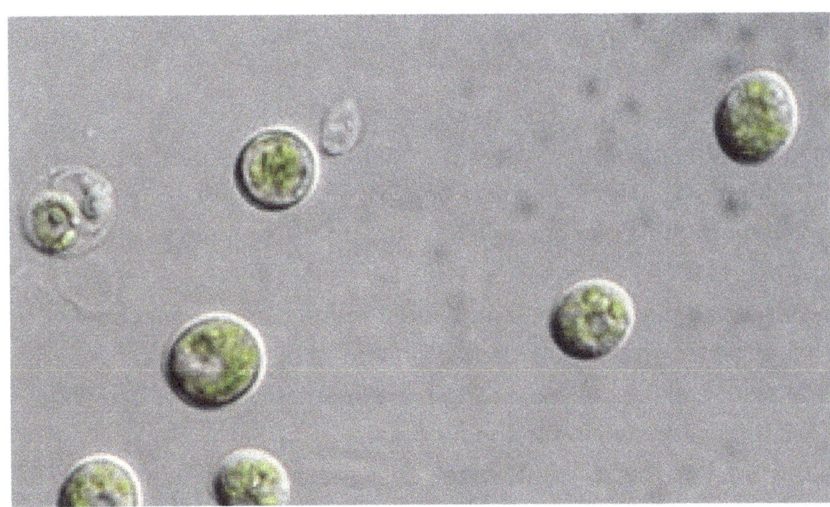

CO_2 is a substrate in photosynthesis and a product of cellular respiration; therefore, in both reactions, we can use CO2 to determine the reaction rate. CO_2 indicators are also pH indicators. How does that work? The following reactions show what happens when CO_2 dissolves in water:

$$CO_2 + H_2O \leftrightarrow H_2CO_3 \leftrightarrow HCO_3^- + H^+$$

Note that when dissolved, CO2 forms bicarbonate, which dissociates into ions. The key ion here, of course, is H+ as the concentration of H+ determines the pH of a solution. The more H+ there is in a solution, the more acidic the solution becomes, and the fewer H+ there are, the more basic the solution becomes. The pH indicator for this experiment gradually changes from yellow at low pH to purple at high pH.

Q1. Why is it possible to use a pH indicator to measure the rates of photosynthesis and cellular respiration?

A. Preparing the Experimental System

Instructions
1. Label the three cuvettes as follows:

Note: Avoid the area where the spectrophotometer will shine a light beam through your sample.

2. Add 2 mL of water to the B cuvette.
3. Add 2 mL of CO_2 indicator to the C cuvette.
4. Add about 10 ml of distilled water to a glass beaker.
5. Remove storage fluid from the cuvette containing algal beads with a plastic pipette. Place storage fluid in the plastic waste beaker.
6. Add 1 ml of distilled water to the algae beads and incubate for 5 min to wash the indicator solution off the beads.
7. Remove distilled water with a plastic pipette and add it to the plastic waste beaker.
8. Cut the tip off of a pipette to form a scoop (as demonstrated by your instructor).
9. Use the scoop to transfer 10 algae beads to each experimental cuvette (L and D).
10. Add 2 mL of fresh CO_2 indicator solution to the A cuvette.
11. Place a sheet of aluminum foil over all the cuvettes until you are ready to start the experiment.

Using Cuvettes in a Spectrophotometer
12. Wipe the outside of each cuvette with a lint-free tissue.

13. Handle cuvettes only by the top edge of the ribbed sides.

14. Dislodge any bubbles by gently tapping the cuvette on a hard surface.

15. Always position the cuvette so the light passes through the clear sides.

B. Determining the Rate of Photosynthesis and Cellular Respiration

Instructions

1. Use the Indicator Guide to determine the pH of the experimental cuvettes (Time = 0) and record in Table 4.2.

2. Placing the cuvettes on a sheet of white paper will make it easier to determine pH by color.

3. Launch the Spectral Analysis application.
    a. Follow the instructions to connect your device to the spectrophotometer.
    b. Click on Absorbance and then select "vs. Concentration." Allow the spectrophotometer to warm up for approximately 90 seconds.
    c. Calibrate the spectrophotometer by placing the B cuvette into the spectrophotometer and selecting "Finish Calibration." Remove the B cuvette from the spectrophotometer.
    d. In the table, change the name of the concentration column to "Duration" and the units to "Minutes."
    e. Place the C cuvette into the spectrophotometer. Enter the wavelength of 550 nm into the textbox provided. Click "Done."

4. Take initial measurements:
    a. Gently invert cuvette A and compare it to the indicator guide to determine its pH.
    b. Remove the C cuvette from the spectrophotometer and replace it with cuvette A.
    c. Wait 10 seconds for the readings to stabilize.
    d. Click "Keep."
    e. Enter 0 minutes into the time textbox.
    f. Record the A_{550} (Time = 0) for the cuvette in Table 4.2.
    g. Lay cuvette A on its side 15-25 Cm under a desk lamp fitted with a 1600 lumen bulb. Make sure the algae beads form a single layer.

Table 4.2. Change in pH and Absorbance Due to Photosynthesis and Cellular Respiration

Time (min)	Light			Dark		
	Indicator color	pH	A_{550}	Indicator color	pH	A_{550}
0						
5						
10						

15						
20						
25						
30						

5. Repeat step 4 to take measurements every 5 minutes for 30 minutes. Be sure to record the pH based on indicator color and gently invert the cuvette before each measurement.
6. After 30 minutes, click "Stop" and remove the cuvette from the spectrophotometer.
7. In the table on the Spectral Analysis App, change the name of Dataset 1 to "Light."
8. Calibrate the spectrophotometer by adding the B cuvette and clicking "Calibrate Spectrophotometer." Then select "Finish Calibration."
9. Click the "Collect" button and select "Create New Dataset."
10. Create a foil "coz-ee" for cuvette A (should fully cover the cuvette but should easily slide off and on for measurements).
11. Repeat step 4 every 5 minutes for 30 minutes, recording your results in Table 4.2.
12. Replace the cuvette in the foil coz-ee and under the desk lamp between measurements.
13. Gently invert the cuvette before each measurement.
14. Compare and record the indicator color and record the A_{550} at each time interval.
15. After 30 minutes, click "Stop" and remove the cuvette from the spectrophotometer.
16. In the table, change the name of Dataset 2 in the Spectral Analysis App to "Dark."
17. Use the Spectral Analysis app menu to save your data for further analysis.

C. Calculations and Analysis

Instructions
1. Use the tools provided by the Spectral Analysis App to apply a linear curve fit to your graph.
2. Choose two points on each line to use to determine the slope of the line.
3. Record the A_{550} and time of each point in Table 4.3.
4. Subtract the later data point from the earlier data point for both variables to determine the change (one will be negative!).
5. Divide the change in A_{550} by the change in time to calculate the slope.

	A$_{550}$ (from graph)	Time (from graph)	Δ A$_{550}$ Δ Time
Linitial			
Lfinal			
Lf – Li			
Dinital			
Dfinal			
Df – Di			

Q2. Why does the A$_{550}$ of the solution change in your cuvette?

Q3. What does the slope of each line on your graph represent?

Q4. Are the slopes of the lines positive or negative? What does a positive slope indicate vs. a negative slope?

Q5. Which sample changed more quickly? Why?

Exercise 4.2. Photosynthetic Pigments

Chromatography is a process that utilizes a liquid (called a solvent) that travels up a solid (like paper) to separate molecules based on their chemical properties. The solvent used in paper chromatography (the type of chromatography we are using) is made of non-polar liquids (acetone and petroleum ether) while the paper (made of cellulose) is polar.

The key to this process is the fact that polar attracts polar and nonpolar attracts nonpolar. Substances that are more polar will bind more strongly to the chromatography paper while non-polar substances will dissolve in the solvent and move up the paper through capillary action. Molecules exhibit different degrees of polarity.

In chromatography, the least polar (and most non-polar molecule) will dissolve in the solvent first and, therefore, will travel the furthest up the paper. The final product of this process is a series of bands with the most non-polar substance concentrated in a band near the top and the most polar substance concentrated in a band at the bottom. Other substances will be spread out in bands in between based on their relative polarity (Figure 4.2).

Figure 4.2. Diagram of a paper chromatogram.

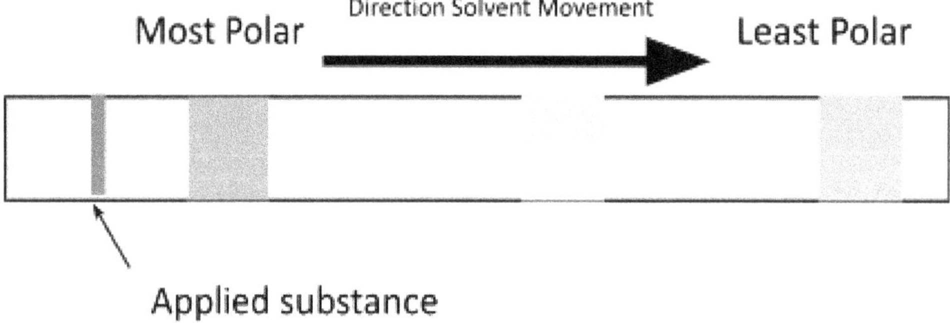

A. Predicting the Chromatogram of a Plant Leaf

We can use the molecular structure of plant pigments to predict what the chromatogram of a plant leaf will look like. In general, leaves contain the pigments chlorophyll a and b, beta carotene, and xanthophyll.

Figure 4.3 displays the molecular structure of each of these molecules. As a rule of thumb, you can determine the relative polarity of a set of molecules by counting the number of polar oxygen atoms in the molecule (more polar oxygen atoms = more polar molecule).

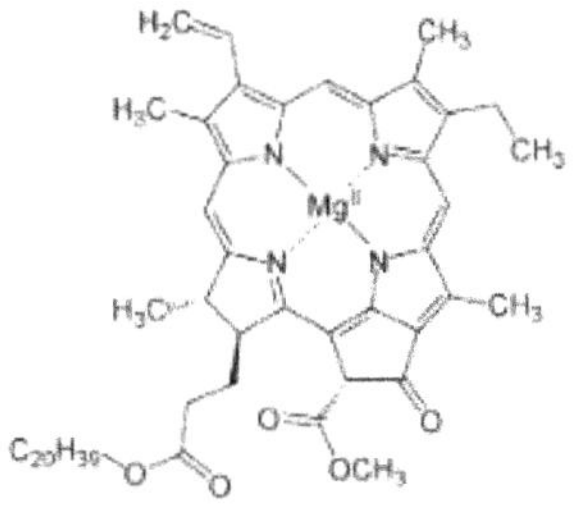

Chlorophyll a

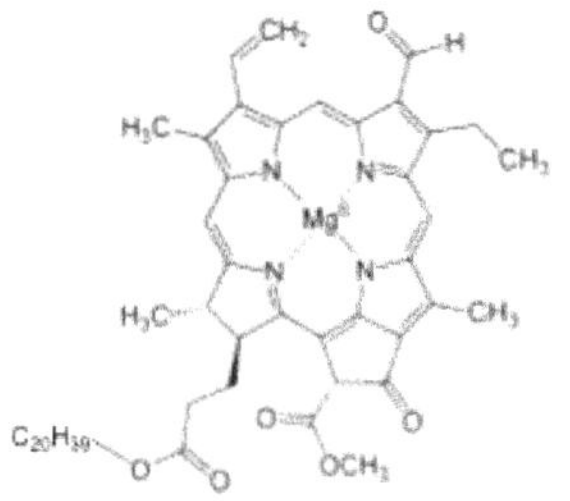

Chlorophyll b

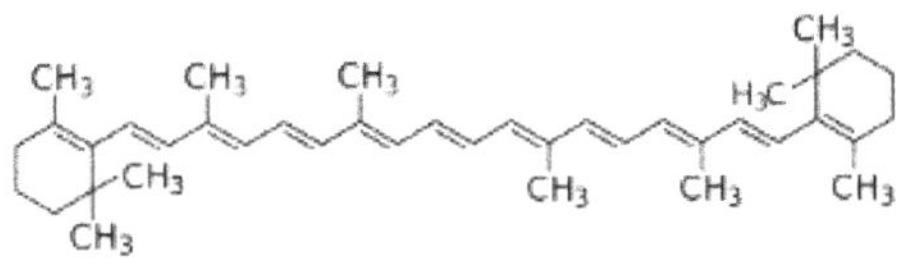

Beta Carotene

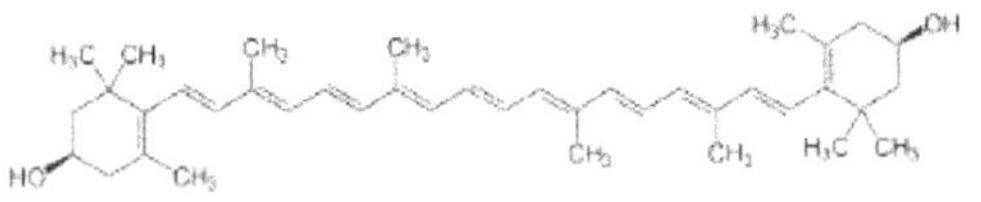

Xanthophyll

Figure 4.3. Molecular structures of plant pigments. Chlorophyll a (A.), Chlorophyll b (B.), Beta-carotene (C.), and Xanthophyll (D.)

Q6. What question is being asked in this experiment?

Q7. What is your hypothesis?

Q8. Complete Figure 4.4 by adding labeled bands to the chromatogram in the order you predicted based on pigment polarity.

Figure 4.4. Predicted results of chromatography of spinach leaf extracts. Complete the diagram by adding bands for ∫-carotenechlorophyll a and b, and xanthophyll.

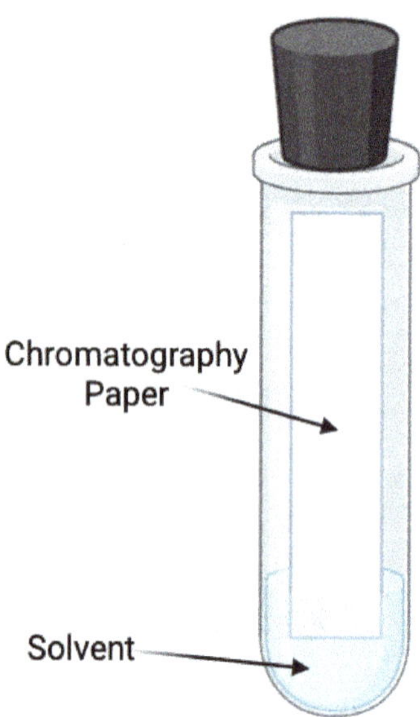

Instructions

1. Obtain a piece of chromatography paper, a few spinach leaves, and a quarter or a dissection probe. Check that the paper strip is not too long for your tube. Use scissors to adjust the length as necessary.
2. Use a pencil to draw a line across the bottom of the chromatography paper approximately 3 cm from the end of the paper.
3. Roll the spinach leaf into a thin "cigar" shape.
4. Place the roll of leaves over the pencil line you drew on the chromatography paper.
5. Roll the edge of the quarter or the tip of the dissecting probe back and forth (with force) over the line a few times.
6. Allow the paper to dry for 1-2 minutes.
7. Move the leaf to an uncrushed region over the same pencil line and repeat.
8. Continue this process until there is a dark line of green over the pencil line.
9. Place the paper into the chromatography tube containing solvent. The level of solvent in the tube should be below the line of pigment. If the level of solvent needs to be adjusted, consult with your instructor
10. Place the stopper in the tube.
11. Watch over the next 15 min as the pigments separate on the paper. Stop before the top band of color moves off the paper.
12. When the topmost color (yellow) gets near the top of the paper, use forceps to remove it from the jar, close the jar, and allow the paper to dry.
13. Use colored pencils to draw your chromatograph in your lab notebook. Label the pigments in your sketch.

Q9. What makes an Oxygen atom polar?

Q10. Why is it important to make sure that you draw a line in pencil at the bottom of the
chromatography paper before adding the leaf extract?

Q11. Did your chromatograph match your prediction? Why or why not?

Exercise 4.3. What Factors Affect the Rate of Photosynthesis & Cellular Respiration?

In this exercise, you will design an experiment that will alter the rates of photosynthesis and
cellular respiration using the same procedure you used in Exercise 4.2. The list below displays
the options that are readily available. Other options may be available, so ask your instructor if
there is something specific you would like to try.

A. Designing an Experiment

- ☐ Determine as a class what independent variable(s) your class would like to test.
- ☐ Each lab group in the class can only perform ONE experimental treatment.
- ☐ You should replace the CO_2 indicator in each cuvette before starting a new experiment.
- ☐ A dark control cuvette is not required for this experiment.

Table 4.4. Available Conditions

Water baths set at different temperatures	Red, Green, and Blue Cellophane
Ice	Aquatic organism
Lamp with 840 lumen bulb	

Q12. What question is your group asking?

Q13. What is your hypothesis? Make sure you answer the question above.

Q14. What do you think will happen in this experiment if your hypothesis is valid?

1. List your procedure step by step on a dry-erase board.
    a. Use Exercise 4.1 as a template to write out the step-by-step procedure you plan to perform.
2. Design a data table on a piece of graph paper to gather your results.
    a. Use Tables 4.2 and 4.3 as templates.

Instructions

1. Run your experiment using the Spectral Analysis App as described in Exercise 4.1 Procedure B.
2. Record your data in your data table.
3. Use the Spectral Analysis App to perform the calculations and analysis outlined in Exercise 4.1 Procedure C.
4. Compile the results of all groups involved in the experiment in a data table and create a graph that optimally displays the data.
    a. Each student should create their own graph.
5. Graphs produced electronically are permitted if they are imported into a word processing program so a figure legend can be added.

6. Figures should include a complete figure legend.
7. Follow all the graphing requirements laid out in the Scientific Writing Handout
8. Prepare a scan of the data table and graph for this experiment for the lab quiz.

Q15. What conclusion can you draw from your data regarding how your chosen independent variable affected the rate of photosynthesis and cellular respiration?

Q16. Does this data support your hypothesis? Why or why not?

4   Applying What You've Learned

A selection of these questions will appear on your lab quiz. Lab quizzes are open book. Type up answers to these questions in advance so that you can copy and paste the answers into the quiz.

1. Explain how the experiments in Lab 5 with algae beads demonstrated that photosynthetic organisms have both chloroplasts and mitochondria.

2. Plant cells in the light perform both photosynthesis and cellular respiration. How does the rate of cellular respiration affect the rate of photosynthesis? Explain your reasoning.

4

5 | Absorption Pre-Lab

Instructions

- ❑ Read the lab manual and then follow the instructions to complete the assignment.
- ❑ You may need your textbook and other resources to complete this assignment.
- ❑ Pre-labs must be completed before the start of the lab.

1. Complete Table 5.1 by listing substances that must cross the plasma membrane of a cell by simple or facilitated diffusion.

Table 5.1. The Role of Each Substance in Maintaining Homeostasis

Substance	Simple/ Facilitated Diffusion?	Role in Homeostasis

2. Differentiate between an internal exchange surface and an external exchange surface.

3. Provide an example of each type of exchange surface in the human body.

4. Do plants have both internal and external exchange surfaces? Why or why not?

5. List at least 3 human organ systems that require an external exchange surface.

5

6. List the two external exchange surfaces of plants.

*Based on Westrich & Berg (2011) Villi, Villi Everywhere: Biological Structures, Surface Area, & Proportional Thinking, American Biology Teacher 73 (3): 156–161.

5 Absorption

While life exists within the plasma membrane, everything it needs to maintain homeostasis resides outside the organism. As a result, the metabolism of any cell or organism is limited by the **efficiency of absorption**, i.e., the amount of matter that enters an organism over a given period. As organisms increase in size and cell number, the demands of metabolism quickly outweigh the absorption abilities of cells. There just isn't enough surface area to get everything an organism needs across that membrane fast enough.

In terms of the evolutionary process, this limitation acts as a strong selective pressure against structures with low surface area. Increased surface area allows more solutes to cross through an exchange surface at any given moment, but the rate of absorption is also impacted by a variety of other factors.

Most **exchange surfaces** are very thin to shorten the distance required for diffusion. Exchange surfaces are often closely associated with internal transport systems to ensure that absorbed material can be rapidly transported to where they are needed. In addition, systems that include exchange surfaces are usually arranged to maximize the concentration gradient.

Finally, solutes must dissolve in water to cross the plasma membrane, so all exchange surfaces must be kept moist. In Lab 5, we will examine how the structure and function of exchange surfaces are key to homeostasis in multicellular organisms.

By the end of this lab, you should be able to:

- ❑ Estimate the surface area of biological structures.
- ❑ List the primary substances that an organism must absorb in order to maintain homeostasis.
- ❑ Describe the factors that most influence the rate of absorption in multicellular organisms.
- ❑ Discuss the importance of high surface area in relation to homeostasis.
- ❑ Compare and contrast the structures that plants and animals use for absorption.

Exercise 5.1. Modeling & Measuring Surface Area*

The skin may be the largest organ in your body, but it doesn't win for surface area! The lungs take the prize there, with twice the surface area lining the digestive system and 30x the surface area of the skin. The amount of surface area of exchange surfaces is a great way to judge their importance to life.

A. Creating an Intestinal Villi Model

Before delving deeper into what these surfaces look like and how they function, we will create a model similar to the one shown here (Figure 5.1) and use simple calculations to determine how changes in structure can impact surface area.

Working models are a great way to simplify complex concepts, especially when we can use math to back up our observations.

Figure 5.1. Model of intestinal villi from Westrich & Berg (2011)

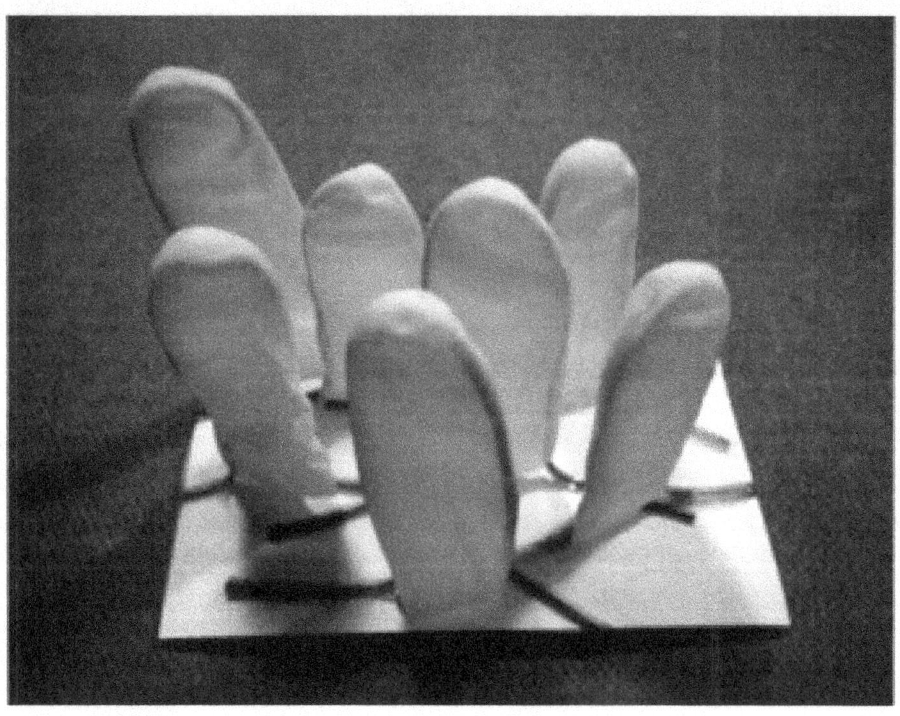

Surface Area Formulas

- ❏ L X W = surface area (mm^2)
- ❏ πr^2 = surface area of a circle (mm^2)
- ❏ $4\pi r^2$ = surface area of a sphere

Instructions

1. Compare the example provided in Figure 5.1 to the index card provided. How many times more surface area do you think the model has than the index card alone?
2. Record your prediction in Table 5.2.
3. Use a ruler and the appropriate calculations to determine the surface area of the index card. Add your calculation to Table 5.2.
4. Cut the fingers off of a disposable glove. Ensure that they are all about the same length.
5. Find the surface area of one villus:
6. Flatten the longest glove finger on a hard surface and use a ruler to measure the length and width of the rectangle formed by the lower part of the longest finger (Figure 5.2). Use the measurements to determine the surface area of that part of the finger.

Note: Remember that the glove has two sides, so double the width.

7. Measure from the top of the rectangle to the tip of the finger to determine the radius of the tip (Figure 5.2). Calculate the surface area of the tip of the figure using the formula for the surface area of a circle.
   a. Add the surface area of the rectangle to that of the circle to obtain the total surface area of the finger.

Note: Each finger half has a half-circle.

Figure 5.2. Determining the surface area of the glove finger from Westrich & Berg (2011).

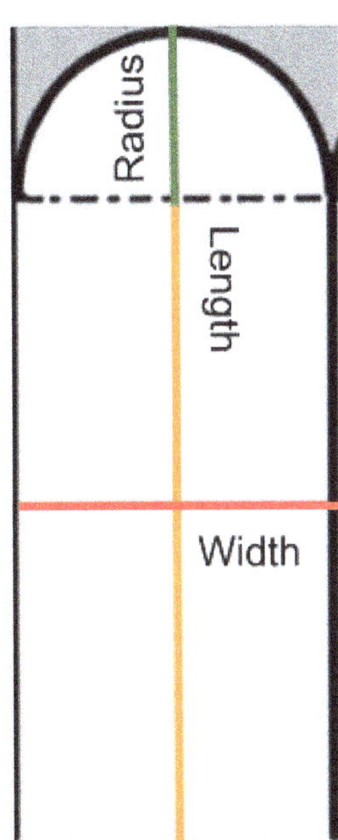

1. Add the surface area of the rectangle to that of the circle to obtain the total surface area of the finger.
    a. Record your calculations in Table 5.2.
2. Use the fingers from the disposable glove (create more as necessary) to create a model of a section of the intestine.
    a. Insert pipe cleaners into the fingers to prop up the villi.
    b. Tape villi along one short edge and one long edge of the index card.
    c. The edges of villi can touch but should not overlap.
    d. Use the number of fingers on each edge to estimate how many fingers would be needed to cover the entire index card.
3. Count the number of villi in your model and multiply by the surface area of one villus.
    a. Record the surface area of your model in Table 5.2.
4. Divide the actual surface area of the model by the surface area of the index card to determine how much more surface area is present in the model.

Q1. What does the surface area of a villus represent in terms of the function of the digestive system?

Q2. Why does the digestive system need high levels of surface area?

Q3. What type(s) of transport is/are used to absorb nutrients?

Q4. How close was your prediction to the actual increase in surface area observed in your model? Explain the differences and similarities.

Table 5.2. Surface Area (SA) Predictions and Measurements for the Model of Intestinal Villi

	Measured SA (mm²)	Predicted Increase in SA when villi are added (? times bigger)
Index Card		
One Villus		
		Actual Increase in SA in Model
Model		

Exercise 5.2. Nutrient Absorption

All organisms must obtain matter from the environment. In plants, this means gases and minerals from the soil. Animals consume more complex matter and break it down into less complex molecules that can be absorbed through the lining of the digestive system. In this exercise, we compare and contrast the nutrient uptake systems of plants and animals. Keep an eye out for other features that make nutrient absorption more efficient.

A. The Human Digestive System

Although we only modeled the villi, the intestine of humans actually contains several levels of structure to increase the efficiency of nutrient absorption. Let's take a closer look!

Instructions

1. Examine and record observations of each of the following:
   a. **Dissected Pig Stomach:** Note the ridges lining the stomach. Why might they be here? Do similar structures exist in other parts of the digestive system?
   b. **Model of Human Digestive System:** What do you notice about the size of the intestine (diameter and length)? Do either of these affect the efficiency of absorption?
   c. **Cross Section of Human Intestine:** The Columnar Epithelium slide from student slide tray. Identify the villi lining the center of the intestine. Look closely at a single cell on the edge of a villus to find microvilli; each one is around 1 micrometer in diameter. The microvilli form a border that looks like a brush extending into the lumen of the intestine.
2. Create a scaled drawing of the intestine cross-section on graph paper and label one villus and the brush border formed by microvilli
3. Create a scaled drawing of a single epithelial cell in graph paper and label the microvilli.

4. Highlight the structures in Figure 5.3 that contribute to the efficient absorption of nutrients in the human digestive system.

Figure 5.3. Diagram of the human digestive system (left), a cross-section of the intestine, and an up-close look at an intestinal villus (right).

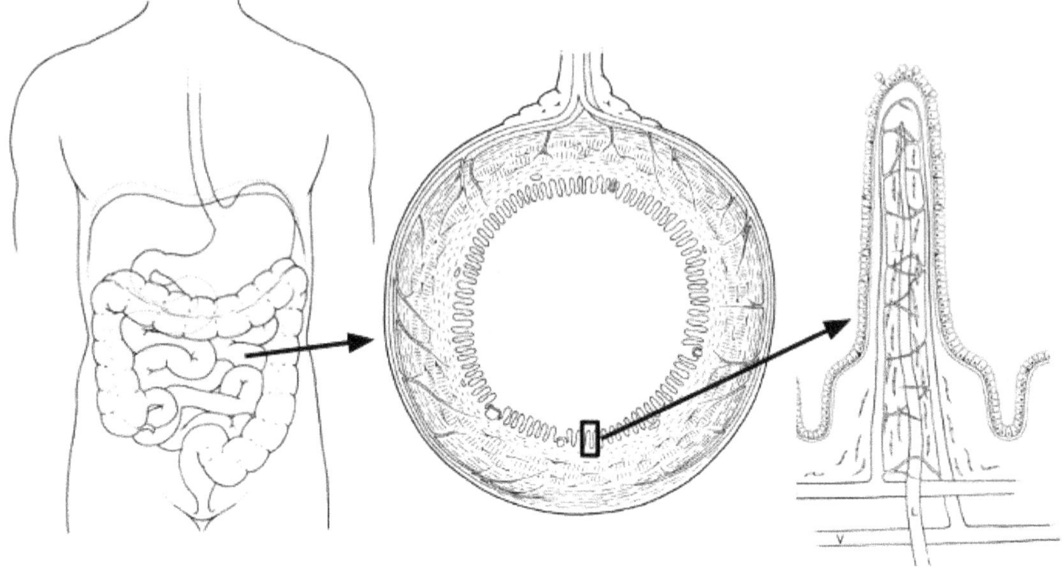

Q5. Describe four structural features that increase the surface area for absorption in the intestine.

Q6. What other features of the intestine contribute to the rapid diffusion and absorption of nutrients?

B. Plant Roots

Plants absorb gases through their leaves and water and minerals through their roots. As plants increase in size, they require more and more nutrients to maintain their tissues, but a plant cannot move to a new place when they run out of minerals. Instead, plants use their roots and interactions with fungi (mycorrhizae) to increase their reach and absorb as much as they can as fast as they can.

1. Use a dissecting microscope to examine a dish of newly germinated radish seeds. Focus on the radicle (young root).
2. Create a scaled drawing of the seedling on graph paper. Label the radicle and the root hairs.
3. Gently separate the soil from the sunflower plant provided to your group. Wash the sample in the dishpan provided. Place the plant on a paper towel to examine.
4. Create a scaled drawing of the root structure on graph paper. Label any structures you see (lateral roots, root hairs, tap root, etc.).

> Note: If you can see individual structures with your eyes, they are NOT root hairs. With your eyes, root hairs look like fuzz.

Q7. What is the difference between a lateral root and a root hair?

Q8. Why does the radicle of a newly germinated seed contain so many root hairs?

Q9. What features create surface area in the roots of a plant?

Q10. Why might a symbiotic relationship with a fungus increase the surface area for absorption in plants?

Exercise 5.3. Gas Exchange Surfaces

Gas exchange is another key process in maintaining homeostasis. All photosynthetic organisms absorb carbon dioxide (CO_2) and release the oxygen gas (O_2) that aerobic organisms require for cellular respiration. Aerobic organisms absorb O_2 as a substrate for cellular respiration and release CO_2 as a product. Gas exchange, as a result, is important to life's ability to store energy and make ATP.

A. Leaf Surface Area vs # Stomata

The amount of light energy a plant can capture and store with photosynthesis is limited to the amount of CO_2 available to plant cells. In order to exchange gases, plant leaves have microscopic pores (stomata) that open and close in response to environmental stimuli.

When stomata are open, gas exchange can occur, but the plant pays a price in the form of water loss through transpiration. The leaves and stomata of plants have evolved a variety of features aimed at both increasing the efficiency of gas exchange and decreasing water loss.

Instructions

1. Use a razor blade and a ruler to cut a square of Zebrina leaf that is approximately 1 cm X 1 cm.
    a. Place the square of leaf tissue purple side up on the stage of a stereomicroscope and cover with a counting card (the hole in the counting card is 4.5 mm²).
    b. Orient the hole so that you can see leaf tissue through the hole using the oculars of the dissecting microscope.
2. Increase the magnification of the dissecting scope until you can see stomata.
3. Count the number of stomata on the bottom (purple side) of the square of tissue. Record the data in the "Number on bottom" column of Table 5.3.
4. Turn the tissue over, replace the counting card, and count the number of stomata on the top of the leaf. Record the data in the "Number on top" column of Table 5.3.
5. Each group should count the top and bottom of two leaves.
6. Complete Table 5.3 by determining the number of stomata per mm²:
    a. Use the data gathered by your group to determine the average number of stomata on the top and on the bottom of a Zebrina leaf.
    b. Divide the average number of stomata by the area of the square in the counting card (the hole in the counting card is 4.5 mm X 4.5 mm, or 20.25 mm²).

Table 5.3. Comparison of Stomatal Frequency on the Leaf Top and Bottom Epidermis of Zebrina

	# stomata on the bottom of the leaf	# stomata on top of the leaf
Leaf 1		
Leaf 2		
Leaf 3		

Leaf 4		
Average number of stomata		
Area of counted region (mm²)	20.25	20.25
Average # stomata /mm²		

Q11. How does the frequency of stomata on top of the leaf differ from the frequency on the bottom of the leaf?

Q12. Why might the top and bottom of the leaf have different stomatal frequencies?

Q13. Would different species of plants have different stomatal frequencies? Why or why not?

B. Gills

Gill is the term used for the gas exchange structures of many organisms, including fish, mollusks, and crustaceans. The gills of different organisms are analogous, as, despite their distinct evolutionary histories, all types of gills exhibit the same features: multiple flaps of thin feathery tissue.

In fish, the gills appear red because they are full of tiny capillaries carrying O_2-poor blood to the gills and O_2-rich blood back to the body. The rate of ATP production in aerobic life forms is directly related to the availability of O_2.

As a result, the more complex the gas exchange surfaces an organism exhibits, the more energy that organism has.

In Exercise 5.3, we will explore the relationship between gill surface area and swim speed in fish.

Instructions

1. Examine the demonstration of fish gills provided by your instructor.
2. Use Figure 5.4 as a guide.

Figure 5.4. Generalized diagram of the structure of fish gills.

3. Sketch and or make observations that show how fish gills exhibit surface area.
4. Examine a small section of the gill with a dissecting microscope. Add any additional evidence of increased surface area to your sketch/observations.
5. Obtain the Fish Gill Data handout from your instructor. Use the data provided to answer the following questions:

Q14. Does a correlation exist between swim speed and gill surface area? Explain why or why not.

Q15. Does a correlation exist between fish mass and gill surface area? Explain why or why not.

C. Lungs

Lung is the term used to describe the gas exchange structure of mammals. As with gills, many analogous structures in the animal kingdom are considered lungs. The lungs of mammals, while drawn as big sacs, are actually made of very finely meshed tissues.

In three dimensions, the tissues look like bunches of grapes. Each "grape" is called an **alveolus,** and your lungs are filled with millions of alveoli to increase the efficiency of gas exchange with the environment.

Instructions

1. Examine a slide of lung tissue with your compound microscope (you will need to go up to high power.) Identify a "cluster" of alveoli.
2. Create a scaled sketch of a cluster of alveoli on graph paper. Label the lumen (where the air goes) and the respiratory surface (exchange surface).
3. Use the formula for the surface area of a sphere to calculate the surface area of a single alveolus. Record your calculations on graph paper.
4. Reduce magnification to 10X and look at the sample as a whole. Think about the surface area of the tiny slice of tissue you are examining.
5. Return to the lung model. Consider how many tiny slices of lung tissue, like the one on your slide, make up a lung.

Q16. Explain how the structure of the human lung provides an efficient surface for gas exchange.

Q17. Lung tissue is extremely fragile. Use what you observed about tissue structure to explain why this might be the case.

5   Applying What You've Learned

A selection of these questions will appear on your lab quiz. Lab quizzes are open-book. Type up answers to these questions in advance so that you can copy and paste the answers into the quiz.

1. Would you expect the dermal tissue of a cactus to contain more or fewer stomata than Zebrina? Explain your reasoning.

2. The breathing organs of terrestrial animals are located within their bodies, while those of fish are located on the body surface. Why are these organs in different locations in terrestrial and aquatic animals?

3. Fish gills, human lungs, and animal intestinal linings all evolved structures to increase the efficiency of absorption. What features do these structures share, and how are those features related to the function of absorption?

<table>
<tr><td>6</td><td>

Internal Transport Mechanisms Pre-Lab

</td></tr>
</table>

Instructions

- ❑ Read the lab manual and then follow the instructions to complete the assignment.
- ❑ You may need your textbook and other resources to complete this assignment.
- ❑ Pre-labs must be completed before the start of the lab.

1. Why can't large multicellular organisms rely on diffusion alone for transport?

2. In plants, negative/positive (circle one) pressure is used to move water, and negative/positive (circle one) pressure is used to move sugars.

3. Compare open and closed circulatory systems. Which would you expect to have higher internal pressure? Why?

4. Sketch a simple diagram of how pressure moves fluid in one of these systems (choose: plant xylem, phloem, insect circulation, or human lungs). Label arrows with "+" for positive pressure or "−" for negative pressure.

6 Internal Transport Mechanisms

As multicellular organisms increase in size, their surface–area–to–volume ratio decreases, making diffusion too slow to deliver oxygen, water, and nutrients to all cells in their bodies. As a result, organisms evolved internal transport systems that use pressure differences to move materials efficiently. This lab explores the differences between positive and negative pressure and how they apply to different internal transport systems.

By the end of this lab, you should be able to:

- ❑ Differentiate between positive and negative pressure.
- ❑ Compare how plants and animals perform internal transport.
- ❑ Relate the structure of a heart to its efficiency as a pump.
- ❑ Describe several ways in which different animals take different approaches to internal transport.

Exercise 6.1. The Mechanics of Internal Transport

The goal of internal transport is to move substances, such as blood, quickly and efficiently to the body's individual cells. While diffusion can be very fast over very short distances, it is very slow over longer distances and, therefore, cannot be used for internal transport.

The primary mechanism of internal transport in multicellular organisms is bulk flow, the movement of a substance from an area of high pressure to an area of low pressure. To understand how this works, we need to begin with an understanding of positive and negative pressure.

A. Positive vs. Negative Pressure

The difference between positive and negative pressure is as simple as pushing versus pulling. Substances move from areas of higher pressure to areas of lower pressure. In positive-pressure systems, a pump, like the heart, is used to create pressure behind a substance, pushing it forward toward areas of lower pressure. In negative-pressure systems, expansion creates an area of very low pressure, allowing higher surrounding pressure to push the substance into that space.

Instructions

1. Obtain a syringe and a beaker of water.
2. To model negative pressure, put the tip of the syringe in the water and pull the plunger.

Q1. What happens to the amount of space inside the syringe when you pull the plunger?

Q2. How does this affect the pressure inside the plunger?

Q3. Why does the water flow into the plunger?

3. To model positive pressure, hold the tip of a water-filled syringe over the beaker of water and push the plunger.

Q4. What happens to the amount of space inside the syringe when you push the plunger?

Q5. How does this affect the pressure inside the plunger?

Q6. Why does the water leave the plunger?

4. Complete Figure 6.1 by labeling the higher- and lower-pressure regions for each experiment.

Figure 6.1. The impact of pulling and pushing the plunger on the pressure in the syringe.

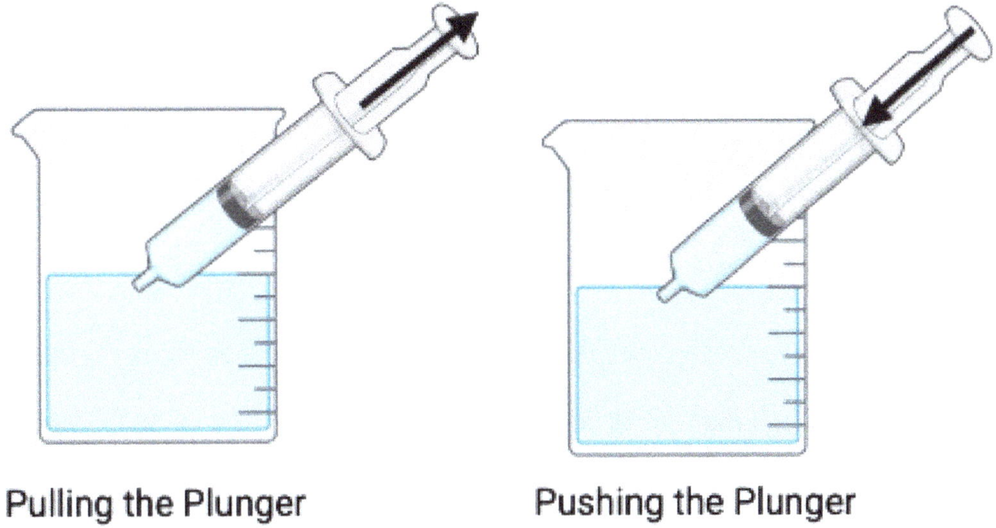

B. Positive vs Negative Breathing

Most mammals use negative pressure to pull air into their lungs and positive pressure to push air out. Recall that negative pressure requires an expansion of space for a substance to flow into. By expanding space, the pressure in that space decreases, making the pressure outside the space higher than inside, so the substance, in this case, air, flows into the space.

In your body, this process is controlled by the diaphragm muscle. When that muscle contracts, it creates space inside your body, and that causes air to flow into your lungs. When the diaphragm relaxes, space decreases, pressure increases, and air moves passively out of the lungs. In this procedure, you will make a model of a lung to simulate inhalation and exhalation.

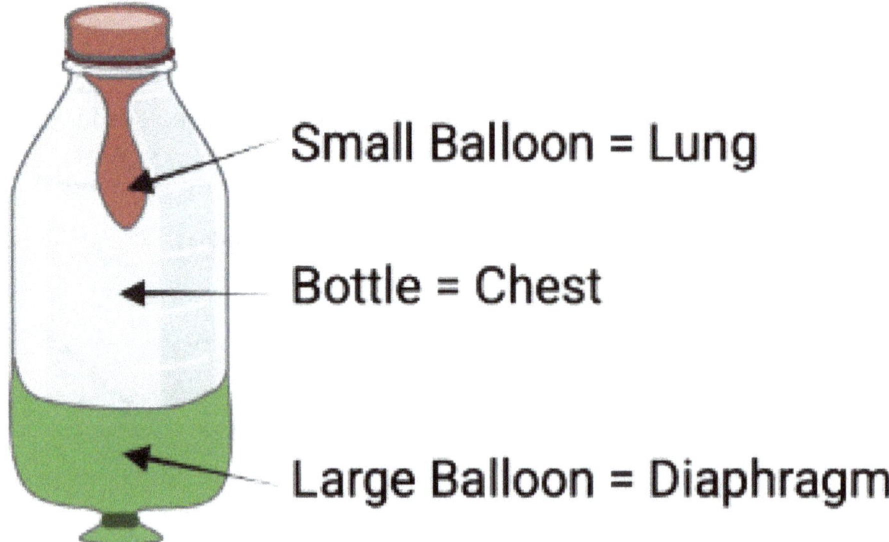

Instructions

1. Create a lung model using a small balloon and a big balloon, and an 8-16 oz plastic water bottle.
   a. Use a knife or scissors to cut off the bottom third of the bottle. Trim the top of the bottle to remove sharp edges.
   b. Use scissors to cut off the top of the large balloon. Tie the other end of the balloon with a knot.
   c. Stretch the large balloon over the opening at the bottom of the water bottle.
   d. Push the top of the small balloon through the neck of the bottle and stretch the end to cover the opening and seal the bottle.
2. Gently pull on the knot of the large balloon. This simulates the contraction. Record your observations in the first row of Table 6.1.
3. Let go of the diaphragm to simulate the relaxation of the diaphragm. Record your observations in the second row of Table 6.1.
4. Push the diaphragm into the bottle. Record your observations in the third row of Table 6.1.

Table 6.1. Results of Manipulating a Balloon and Bottle Model of the Human Lung

Model manipulation	Bottle Volume (increase/ decrease)	Pressure Change (increase/ decrease)	Impact on the small balloon (inflate/deflate)	Is the change active or passive?*
Pull the dia-phragm balloon down				
Release the dia-phragm balloon (after pulling)				
Push the dia-phragm balloon				

*Active means the process requires an input of energy on your part

Q7. Which of the lung model manipulations simulated negative pressure breathing?

Q8. Which of the lung model manipulations simulated positive pressure breathing?

Q9. If the diaphragm in the lung model were punctured (analogous to a collapsed lung), how would ventilation be affected? How does this compare to puncturing the lung tissue itself?

Exercise 6.2. Plant Internal Transport

Plants utilize two separate internal transport mechanisms. In the stem, the cells of the xylem move water against gravity from the root of the plant to its shoot. Also in the stem, the cells of the phloem move sugars from where they are made (**source**) to where they are used (**sink**).

In this exercise, we will determine what types of pressure are involved in plant transport.

A. Plant Stem Structure

The arrangement of cells in the vascular bundle is always the same. Cells of the **xylem** are larger and have thick walls and white centers. They are always located closest to the center of the stem. Just outside of the xylem, there is a band of (usually blurry) tissue known as the **vascular cambium**, a lateral meristem.

The cells of the **phloem** are medium-sized cells that are always adjacent to smaller cells. They are always on the side of the vascular bundle closest to the dermal tissue. The larger cells are **sieve tube members,** while the smaller, square cells are their **companion cells**. Note that the thick-walled cells (usually red) outside of the phloem are ground tissue fibers that support the vascular tissue.

Instructions

1. Obtain the "Monocot and Dicot stem" slide. Use the reference poster provided in the lab to determine which section of the slide depicts the dicot stem.
2. Focus on the dicot stem with the 10X objective on the compound microscope.
3. Move one of the vascular bundles in the stem into the center of the field of view and, after making sure you are in focus, increase to the 40X objective. Re-center the vascular bundle if necessary.
4. Create a scaled sketch of the vascular bundle on graph paper. Label all the terms listed in bold in the description above, and include observations

Q10. Compare the cell wall of the xylem to the cell wall of the sieve tube members in the phloem. What explains the difference in thickness?

Q11. Why are the sieve tube members always associated with companion cells?

B. Water Transport

Is water pushed or pulled up the stem? If water is pushed up by positive pressure in the roots, removing the roots would stop transport. In this exercise, we examine the absorption of water by a stem to determine whether water continues to move upward when roots are absent.

Instructions
1. Add several drops of blue or red food coloring to a beaker of water.
2. Place a stem of celery or a flower (white daisy or carnation) into the beaker.
3. Use a timer to determine how long it takes for the dye to reach the top of the stem.

Q12. How will you know when the dye has reached the top?

Q13. What type of transport tissue would you expect to find the dye in?

4. When the dye has reached the top of the stem, remove it from the beaker
5. Use a razor blade to slice into the stem and then cut a few thin sections of stem.
6. Make a wet mount of the thinnest section you created.
7. Examine the wet mount with the dissecting and compound microscopes.
8. Create a scaled sketch of the section and label the location of the dye.
9. Compare your sketch to the one you drew of a vascular bundle in Exercise 6.1, Procedure A, to identify the xylem and phloem in your stained section.

Q14. How long did it take for the dye to reach the top of the stem? Compare your results with other groups? Provide hypotheses to explain similarities and differences.

Q15. Which type of vascular tissue contains the stain?

Q16. Do your results suggest that water travels up a stem?

Q17. The roots are at one end of plant vascular tissue and stomata are at the other end. If the roots don't push the water up, how do the stomata create the pressure differ-ence that moves water up the stem?

C. Sugar Transport

Plants also transport sugar from source (where it is made) to sink (where it is used). Most of the sugar in plants is made in leaves.

In this experiment, we will use observations of what happens when a leaf is cut off a stem to understand how sugar transport works. To understand this experiment, think about what you have learned so far about positive and negative pressure and answer the following questions:

Q18. If a tube is under negative pressure, do the contents of the tube move toward or away from the rupture?

Q19. If a tube is under positive pressure, do the contents of the tube move toward or away from the rupture?

Instructions

1. Obtain a glucose test strip and a well-watered plant (cucumber, squash, or nasturtium) that has been kept under bright light for a few days.
2. Cut a leaf off the stem of the plant. Liquid from the leaf should exude from the leaf petiole.
3. Tap the exudate from the leaf onto the glucose test strip. Follow the instructions to interpret the results of the glucose test strip.

Q20. What does the fact that the sap comes out of the leaf suggest about the type of pressure in the tissue that holds that solution?

Q21. What was in the solution that came out of the leaf?

Q22. What type of pressure does a plant use to move sugar from its leaf to its various sinks?

Q23. Xylem relies on dead cells, while phloem relies on living cells. Why might plants use different strategies for transporting water versus sugars?

Exercise 6.3. Animal Internal Transport

Animals also require an internal transport mechanism to move nutrients, gases, and wastes from where they are produced or absorbed to where they are needed. Very small animals, such as flatworms, rely on diffusion, whereas larger animals require faster mechanisms to move substances throughout their bodies.

A. Invertebrate Hearts

Not all invertebrates have internal transport systems, but those that do possess simple or specialized hearts that pump fluid around the body (Figure 6.3). Invertebrate circulatory systems are classified into two types: open and closed.

In an **open circulatory system**, the circulating fluid (**hemolymph**) is released into body cavities, where it directly bathes the cells. In a **closed circulatory system**, the fluid (**blood**) remains within vessels, and exchange of oxygen, nutrients, and wastes occurs as substances diffuse across the walls of small vessels into the tissues.

Figure 6.3. Diagram of a closed invertebrate circulatory system. Blue indicates deoxygenated blood and red indicates oxygenated blood. Other colors represent a mixture of oxygenated and deoxygenated blood.

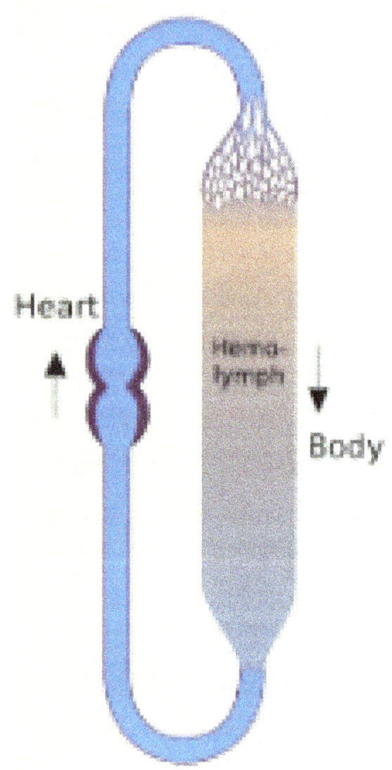

Instructions

1. Create a wet mount of a blackworm.
2. Center and focus the slide until you can see the dorsal blood vessel running down the length of the worm. Increase the magnification until you can clearly see the details around the vessel.
3. Observe the worm for several minutes and note your observations in the appropriate section of Table 6.2.
4. Determine the pulse rate (beats/minute) of the worm using the method we used in lab 2. Record the average pulse rate in the observations section of Table 6.2.
5. Hypothesize as to whether the blackworm has an open or closed circulatory system.
6. Create a bulleted scientific explanation of your hypothesis in Table 6.2.

Table 6.2. Types of Circulatory System Exhibited by a Blackworm

Claim:	
Observations:	Reasoning:

7. Create a wet mount of a *Daphnia* (a crustacean).
8. Center and focus the slide until you can see the pumping heart
9. Observe the pumping of the *Daphnia* heart for several minutes and note your observations in the appropriate section of Table 6.3.
10. Determine the pulse rate (beats/minute) of a Daphnia using a method similar to the one we used for the worm. Record the average pulse rate in the observations section of Table 6.3.
11. Hypothesize as to whether the blackworm has an open or closed circulatory system.
12. Create a bulleted scientific explanation of your hypothesis in Table 6.3.

Table 6.3. Type of Circulatory System Exhibited by Daphnia

Claim:	
Observations:	Reasoning:

Q24. What feature(s) were most important in determining whether these organisms had open or closed circulatory systems?

Q25. How does having a closed circulatory system impact the efficiency of internal transport in invertebrates?

Q26. How does the type of circulatory system exhibited by an organism affect its lifestyle?

B. Mammalian Heart Anatomy

The heart is the pump of the mammalian circulatory system. The contraction of the heart muscle generates the high pressure that drives blood through the body's blood vessels via bulk flow.

In this exercise, we will dissect the heart of a pig, which is approximately the same size as your own heart and very similar in structure. Use the keys and models provided to identify the vessels and cavities of the heart and determine the direction of blood flow.

Instructions

1. Place the heart on a dissecting tray with the rounded side (base) up and the pointed side (apex) down.
2. Identify the left and right sides.
   a. The **left ventricle** extends to the apex.
   b. The left ventricle will feel firmer and thicker than the **right ventricle.**
   c. Note that the heart's left and right are opposite yours when you face it.
3. Identify the **left and right atria** (atrium is singular).
4. Use the guides and models provided to identify the **pulmonary arteries**, the **pulmonary veins**, the **vena cava** and the **aorta**. Label all structures indicated in bold in the external anatomy diagram in Figure 6.5.
5. Insert scissors into the **right atrium**, cut downward into the right ventricle, and open it like a book.
   a. Observe the thickness of the ventricle wall.
   b. Locate the **valve** between the right atrium and right ventricle.
6. Make a similar cut through the left atrium and left ventricle.
   a. Observe the thickness of the ventricle wall
   b. Locate the valve between the left atrium and left ventricle
7. Use the guides and models provided to identify all structures indicated in bold in these instructions in the internal anatomy diagram in Figure 6.5.

> Note: The heart contains a total of 4 valves. Make sure you find them all!

8. Use a blunt probe to study the function of the valve between the left atrium and ventricle.
   a. These valves are one-way; they let blood move from the atrium to the ventricle, but not the other way.
9. Find the **chordae tendineae** ("heart strings") attached to the valve.
   a. These cords stabilize the valve by preventing it from inverting under high pressure.
10. Gently push your fingers or a blunt probe through the vessels to determine which chambers or vessels they connect to. Use what you learn to trace the flow of blood through the heart.

a. Deoxygenated blood from the body enters the heart through the vena cava.

b. Oxygenated blood exits the heart through the aorta.

c. The pulmonary arteries and veins connect the heart to the lungs.

11. Complete Figure 6.4 by drawing arrows to indicate the direction of blood flow through the heart.

Figure 6.4. External and internal anatomy of the heart.

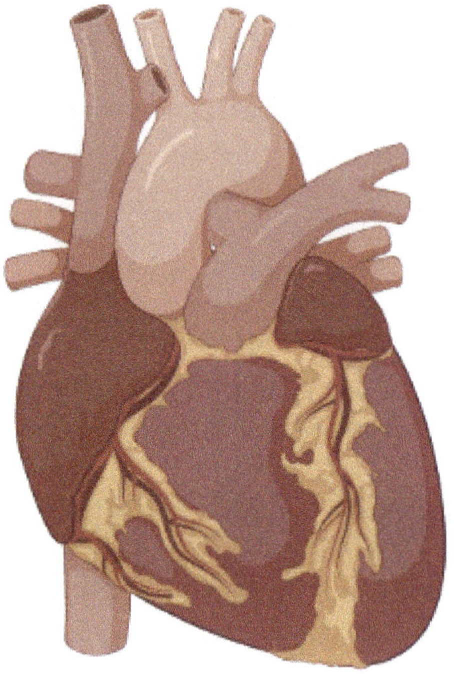
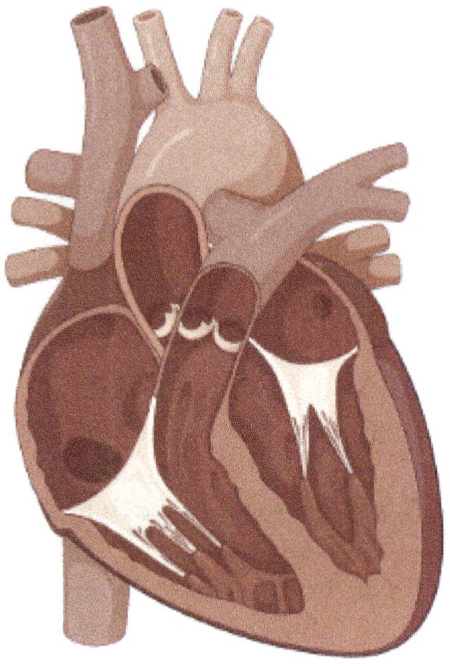

Q27. What is the function of the atria? How does the structure of these chambers reflect
their function?

Q28. What did you note about the thickness of the ventricle wall on the left vs. the right
side of the heart? Does this difference in structure reflect different functions for the
left and right ventricles?

Q29. How does the arrangement of the mammalian heart ensure that oxygenated and
deoxygenated blood never mix?

C. Verterbrate Hearts with Fewer Chambers

Vertebrate hearts evolved over long periods, and by examining the structures of other animals,
we can explore how the heart may have evolved. Fish, amphibians, and mammals/birds have
different heart structures.

Fish have two chambers, a single atrium, and a single ventricle, while amphibians have a
3-chambered heart (2 atria and one ventricle). Mammals and birds have 4-chambers (2 atria
and 2 ventricles).

In this exercise, you will explore the impact of heart structure on the efficiency of internal
transport.

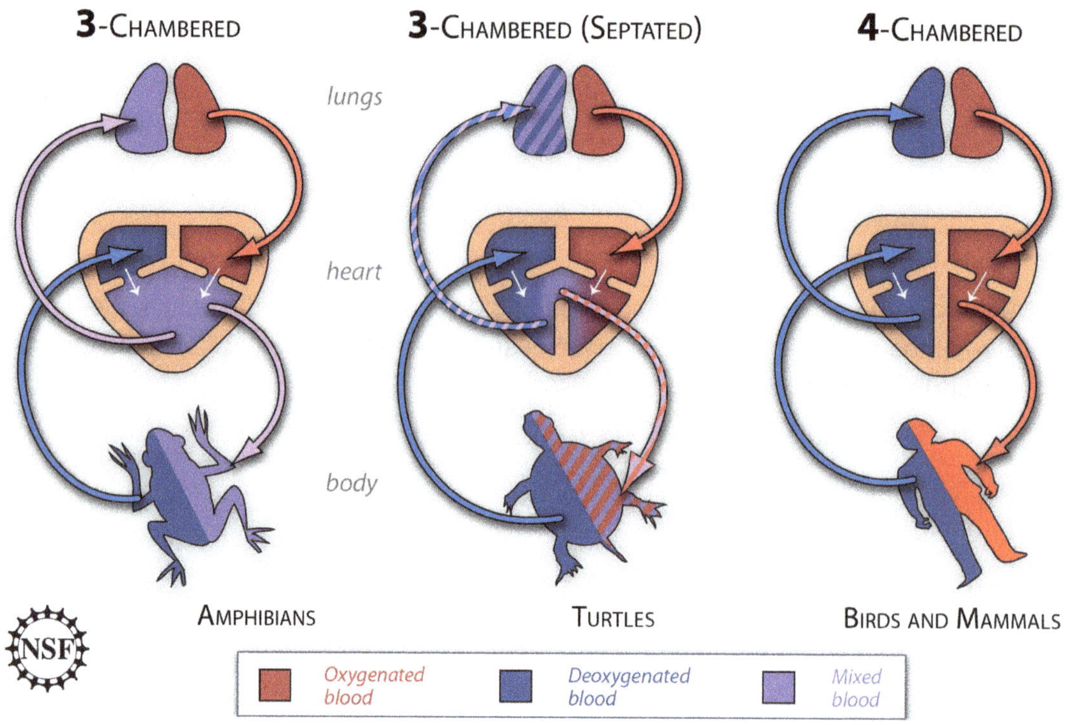

Blue indicates deoxygenated blood and red indicates oxygenated blood. Other colors represent a mixture of oxygenated and deoxygenated blood.

Instructions

1. Use the resources provided by your instructor and Figure 6.6 to complete Table 6.4.

Table 6.4. Comparison of the Circulatory Systems of Different Animals

Heart type	Number of Chambers	Number of Circulation Circuits	Does high O_2 blood mix with low O_2 blood?	Pressure (High/Low/ Moderate)	Example Animal
Fish					
Amphibians					
Bird/ Mammal					

Q30. How does the number of chambers in a heart affect the efficiency of internal transport?

Q31. How does the efficiency of internal transport affect the lifecycle of an organism?

Q32. List at least three evolutionary trends (non-random, directional changes) you see regarding heart evolution. What selective pressures are driving these changes?

Q33. Do larger organisms require more pressure to perform internal transport? How does that relate to the number of chambers in an organism's heart?

6 | Applying What You've Learned

A selection of these questions will appear on your lab quiz. Lab quizzes are open book. Type up answers to these questions in advance so that you can copy and paste the answers into the quiz.

1. If an air cavity is large, would it be more efficient to use positive pressure or negative pressure for ventilation?

2. During a drought, plants close their stomata to reduce water loss. Does this affect transport in the phloem?

3. When dissecting the pig heart, you observed one-way valves. Why is valve function so critical in larger, high-pressure circulatory systems? What might happen if these valves failed?

4. Endotherms maintain constant body temperature, while ectotherms do not. How do differences in heart structure relate to these metabolic strategies?

5. Both plants and animals need to transport resources across their bodies. Compare the "pumps" and "pipes" in the two systems. In what ways are the solutions similar, and in what ways do they differ fundamentally?

7 | Reproduction & Development Pre-Lab

Instructions

- ❑ Read the lab manual and then follow the instructions to complete the assignment.
- ❑ You may need your textbook and other resources to complete this assignment.
- ❑ Pre-labs must be completed before the start of the lab.

1. Complete the diagram of the animal life cycle in Figure 7.1A as follows:
    - ❑ Label the mature individual, the gametes, and the zygote.
    - ❑ Label the arrows to indicate whether mitosis, meiosis, or fertilization was taking place.
    - ❑ Indicate and label the top (above dotted line) and bottom (below dotted line) to indicate which stages of the life cycle are 1n vs. 2n.

Figure 7.1A. Diagram of the animal life cycle (A) and alternation of generations in plants (7.1B). The larger structures contain cells and represent multicellular stages. Single cells are reproductive cells. The dotted line separates the haploid and diploid stages.

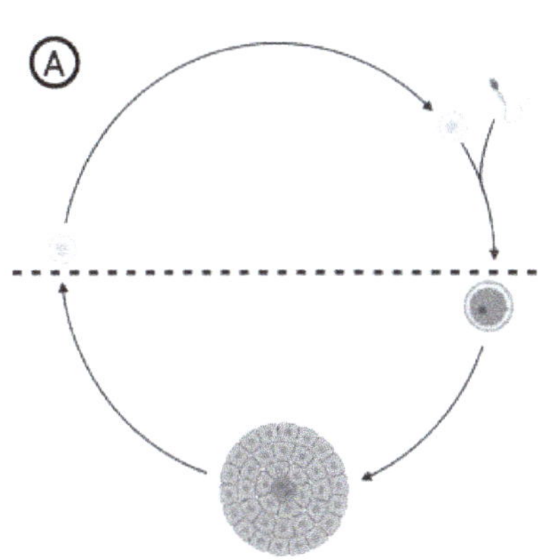

Animal Life Cycle

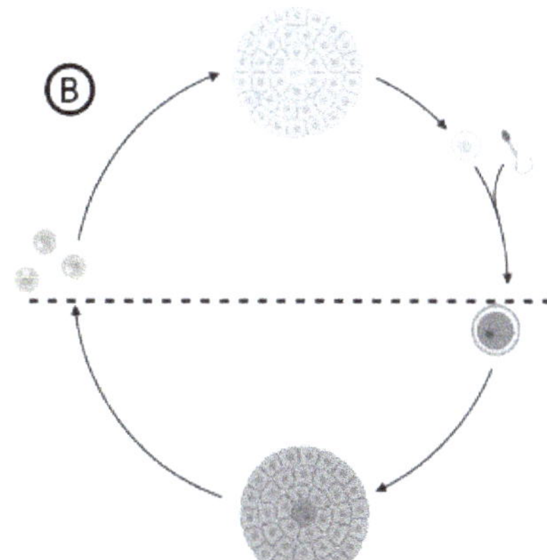

Plant Life Cycle
Alternation of Generations

2. Complete the diagram of the plant life cycle in Figure 7.1B as follows:
   - ❑ Label the sporophyte, gametophyte, spores, gametes, and zygote.
   - ❑ Label the arrows to indicate whether mitosis, meiosis, or fertilization was taking place.
   - ❑ Indicate and label the top (above dotted line) and bottom (below dotted line) to indicate which stages of the life cycle are 1n vs. 2n.

3. What type of cell division is used to produce gametes in plants versus animals?

4. What type of cell division is used to reduce chromosome numbers in plants vs animals?

2. Watch the videos provided for this assignment on the course website to complete Table 7.1.

Table 7.1. Reproductive Strategies in Multicellular Organisms

	Sea Urchins	Frogs	Chickens	Angiosperms
Internal or External Fertilization?				
Quantity of gametes (high, low, moderate)				
Parental Investment (high, low, moderate)				
Survivability of Offspring (high, low, moderate)				

7 Reproduction & Development in Multicellular Organisms

By the end of this lab, you should be able to:

- ❑ Compare and contrast life cycles in plants and animals.
- ❑ Compare the processes of sperm and egg formation in animals to the processes of pollen and ovule formation in plants.
- ❑ Differentiate between internal and external fertilization.
- ❑ Identify differences between early development in plants and animals.
- ❑ Explain how reproduction and life cycles ensure continuity of life.

Exercise 7.1. Gametogenesis

Gametogenesis is the process of producing the sex cell known as a **gamete**. There are two types of gametes: sperm and eggs. Both sperm and eggs are haploid, so when a sperm nucleus fuses to an egg nucleus during fertilization, the result is a diploid cell known as a **zygote**.

A. Gametogenesis in Humans: Oogenesis

All animals use meiosis to produce gametes, but the details vary by species. In the human ovary, each developing egg cell is housed in a fluid-filled sac called a **follicle**. Ovarian germ cells initiate meiosis before birth, but the process halts at prophase I, leaving immature primary oocytes arrested until puberty.

Beginning at sexual maturity, hormones stimulate the development of several follicles each month, though usually only one fully matures. The oocyte in a maturing follicle completes meiosis I with asymmetric cytokinesis, producing a large secondary oocyte and a much smaller polar body, which typically degenerates.

At this point, the secondary oocyte begins meiosis II but arrests at metaphase II. Completion of meiosis II, producing a true ovum and an additional polar body, only occurs if fertilization takes place.

Figure 7.2. Oogenesis. 7.2A. Micrograph of Ovarian Tissue. The circular structures within the tissue are developing follicles. A mature follicle is indicated. 7.2B. The process of meiosis during oogenesis.

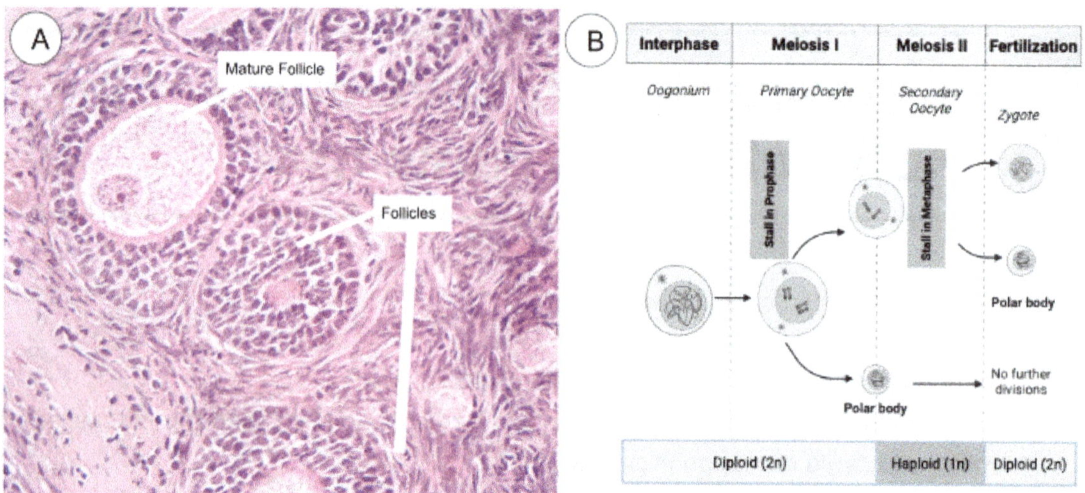

Instructions

1. Obtain the slide "Ovary Cat" to examine with your microscope.
2. At low magnification, locate and center a follicle. Increase magnification to identify stages of oocyte development (Figure 7.2).
3. Create a scaled sketch of a section of the ovary.
4. Use the models and handouts provided by your instructor to locate the different stages of oogenesis.
5. Label an oocyte and primary and secondary follicle stages in your sketch.
6. Under each label, indicate whether the cell has not started meiosis, is at the end of meiosis I, end of meiosis II, or after meiosis.

B. Gametogenesis in Plants

While all plants undergo alteration of generations, the process varies across plant clades. As a rule, gametes are made by gametophytes using mitosis. Recall that, in plants, meiosis is used in the production of spores.

Angiosperms evolved flowers to make **pollination** (transfer of pollen to the carpel) more efficient. The structure of a flower is arranged in concentric rings called **whorls** (Figure 7.3).

Figure 7.3. The whorls of a flower. The dashed line emphasizes the arrangement of floral parts in concentric rings around a central axis.

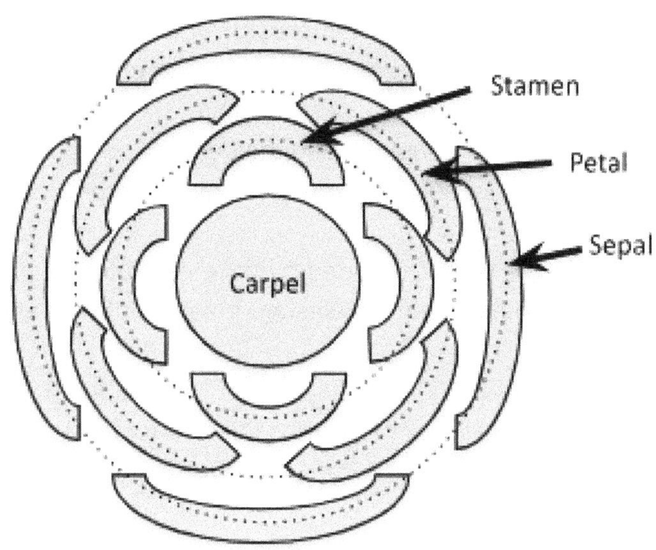

- ❑ The **sepals** function to protect the developing flower bud and, in some flowers, gain a petal-like function in the mature flower.
- ❑ **Petals** have evolved a variety of shapes, colors, and other features to attract pollinators. Flowers that are not pollinated by an animal often lack petals.
- ❑ The **stamen** is the male reproductive structure. Each stamen has a **filament** that positions the anther for pollination, and the anther is the site of pollen formation.
- ❑ The **carpel** is the female reproductive structure. The **stigma** at the top is sticky for capturing pollen. The base is the **ovary**, where female gametes are produced. The stigma and ovary are connected by the **style**, which positions the stigma for efficient pollination.

Gametes are formed in two places in the flower:
- ❑ **Anther:** Meiosis produces spores that undergo mitosis to develop into **pollen grains** (the male gametophyte).
- ❑ **Ovary:** Contains one or more **ovules.** A cell in each ovule undergoes meiosis to produce spores; typically, one spore survives in each ovule, and that spore develops into the female gametophyte, which produces gametes by mitosis.

Instructions

1. Examine a lily flower from the top to identify each of the four whorls shown in Figure 7.3.
2. Make a sketch of the lily from the top on graph paper. Label each whorl.
3. Remove the outer whorl of the flower, the sepals, and make a sketch of one sepal on graph paper
4. Repeat for the next two whorls. Label any parts indicated in the description above.
5. All that should remain of your flower is the carpel. Make a sketch of the carpel and label the parts described above.
6. Examine the demonstration of a dissected carpel provided by your instructor. Locate one or more ovules in the ovary. Estimate how many ovules are present in the ovary of the lily flower.

Q1. Why are polar bodies produced during oogenesis? How does this impact the number of ova a female can produce?

Q2. How many seeds might be produced by a single lily flower? How do you know?

Q3. How does gamete production in plants compare with gametogenesis in animals?

Q4. Students often use the terms "meiosis" and "sexual reproduction" interchangeably. Are they the same thing?

Exercise 7.2. Fertilization Strategies

A. Observing Double Fertilization in Angiosperm

Multicellular organisms have evolved a variety of strategies around fertilization. In general, fertilization either occurs in the environment (**external fertilization**) or within the organism itself (**internal fertilization**).

You completed Table 8.1, which compares reproductive strategies in plants and animals, as part of your pre-lab. If you have time, check your answers with your group before taking a closer look at angiosperm reproduction in this exercise.

Fertilization in angiosperms is called double fertilization because it involves the fusion of sperm to two different cells in the female gametophyte. The female gametophyte consists of approximately eight cells: one egg, one central cell (containing 2 nuclei), and several support cells. One sperm fertilizes the egg to create an embryo, and the second cell fuses with the central cell, creating the endosperm, a source of stored food for the embryo. After fertilization, the ovule develops into a seed containing an embryo, food, and a protective seed coat.

Figure 7.4. Double fertilization and seed development in angiosperms.

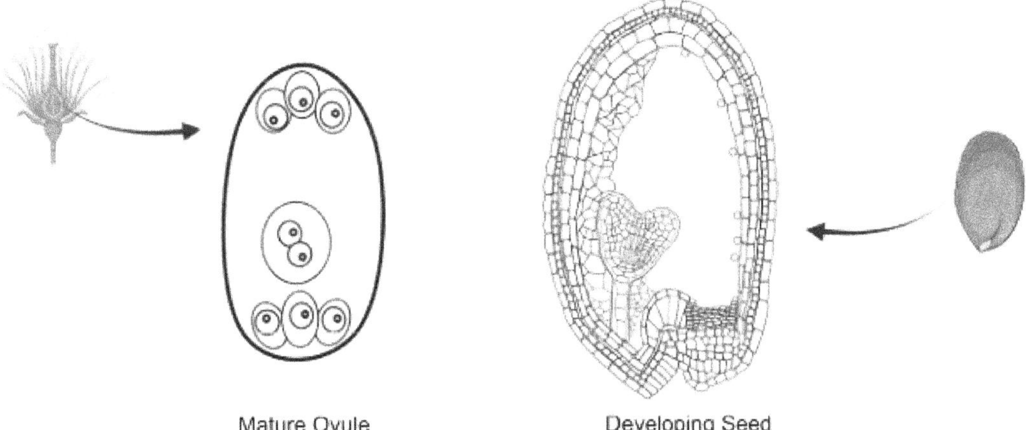

1. Obtain the slide "Lilium Ovary 8 Nucleate Embryo Sac, c.s." to examine with your microscope.
2. Use the image of an ovary cross-section in Figure 7.5 to locate an ovule. You may have to look at several ovules to find one that resembles the mature ovule diagrammed in Figure 7.4.

Figure 7.5. Cross Section of a Lily Ovary. Looking down on a slice of tissue from the region indicated by the line on the flower diagram.

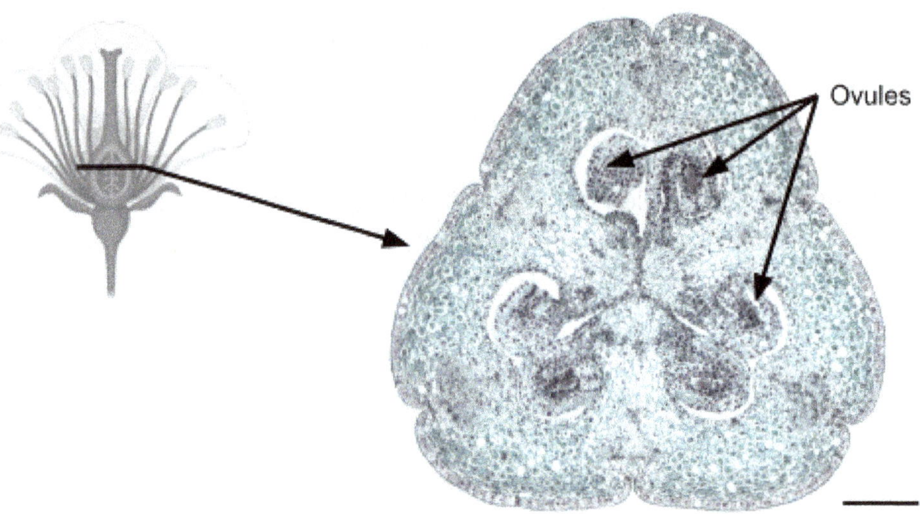

3. Examine the ovule at high power and identify the egg and central cell.
4. Complete the left side of Figure 7.5 by coloring in the two cells in the mature ovule (left) that are fertilized during double fertilization.
5. Work with your group to complete the angiosperms column of Table 7.1.

Q5. Why is angiosperm fertilization called "double fertilization?"

Q6. Is plant fertilization internal or external?

Q7. Which selective pressures favor internal vs. external fertilization in multicellular organisms?

Q8. What is the relationship between the number of gametes produced, parental care, and survivability?

Exercise 7.3. Embryo Development

Almost all organisms begin as a **zygote**, a fertilized egg formed from the fusion of an egg with a sperm. If an organism is multicellular, this single-celled zygote must divide to form the rest of the organism's cells. In addition, the cells must differentiate, that is, they must become different from each other in terms of structure and function.

A. Sea Star Development

All animals start the development process with three common steps.

- ❑ **Cleavage:** cell number increases, but the overall size of the embryo stays nearly the same because there is no cell growth between divisions. Cleavage produces a solid ball of cells called the morula. As development continues, the morula reorganizes into a hollow, fluid-filled ball of cells called the blastula
- ❑ **Gastrulation:** cells push into the embryo at an opening called the blastopore. This process establishes the three primary germ layers—ectoderm, mesoderm, and endoderm—that will later differentiate into all tissues and organs. The resulting three-layered embryo is called the gastrula.
- ❑ **Organogenesis:** organ systems begin to form

In this procedure, we will examine cleavage and gastrulation in prepared slides of sea star embryos. Sea stars, like sea urchins, release their gametes into the environment, allowing us to easily observe the process of development.

All stages of development should be visible in one slide, but there are many embryos on each slide, so you will need to look around to find representatives of each stage.

Instructions

1. Obtain the slide "Starfish Development II Composite, w.m." for examination with your microscope.
2. Use the mid and high-power objectives to identify embryos at each stage of development. Use the resources provided in class to aid in identification.
3. Create a scaled sketch of the following stages: 8-celled, blastula, and gastrula on graph paper.

B. Seed Development & Germination

After fertilization in flowering plants, rapid mitotic divisions transform the zygote into an embryo. Unlike animals, plants do not undergo cleavage, gastrulation, or organogenesis. Instead, both the zygote and the **fertilized central cell** (which becomes endosperm) undergo mitosis.

A crucial milestone in plant development is the formation of **apical meristems**—clusters of undifferentiated cells at both ends of the embryo. These meristems remain active throughout the plant's life, giving rise to all above-ground and below-ground tissues.

Instructions

1. Use the resources provided by your instructor to identify the parts of the developing seed on the right side of Figure 7.6. Label the cotyledons, embryo, endosperm, and seed coat in the developing seed.
2. Connect the cells you colored in in the mature ovule (left side of Figure 7.6) to the structure they develop into in the seed (right side of Figure 7.6).
3. Examine a prepared slide of "Coleus Stem Apex" with your compound microscope. Move to the very end of the shoot to a region that looks a bit like the head of an insect.
4. Use Figure 7.8 to identify the shoot apical meristem, bud primordia, and leaf primordia. The primordia are embryonic structures that develop into buds and leaves. Remember that the stem grows in modular "bud-stem-leaf" units, so you should find leaf and bud primordia repeating this order as you move away from the apex of your specimen.
5. Label the parts of the stem tip on the right side of Figure 7.6.

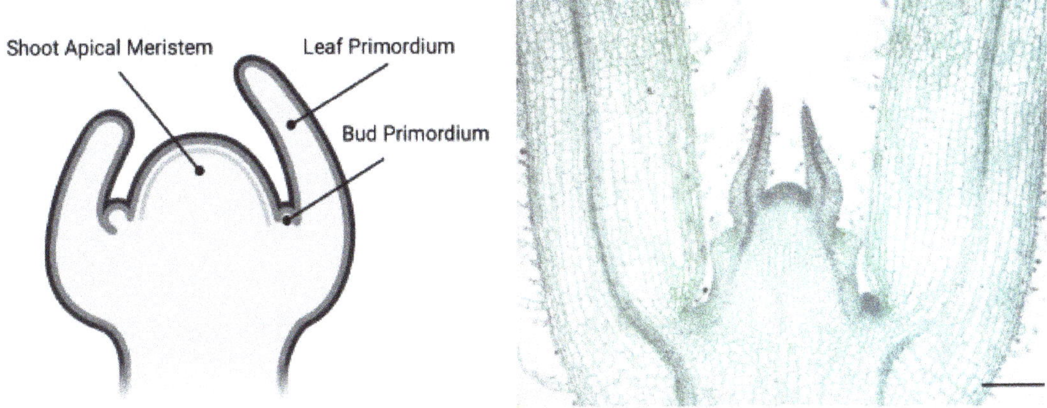

Figure 7.6. The anatomy of the shoot apical meristem. Diagram of a shoot apical meristem (left) and micrograph of Coleus stem tip (right).

Q9. What type of fertilization does the sea star undergo?

Q10. The micrograph on the right side of Figure 7.6 shows lines passing through the leaf primordia. What are these lines?

Q11. Why does early development in all multicellular organisms include so much mitosis?

Q12. In animals, most cells are fully differentiated early in development, but plants maintain meristems throughout their lives. How does this difference affect the way that plants and animals grow?

7 | Applying What You've Learned

A selection of these questions will appear on your lab quiz. Lab quizzes are open book. Type up answers to these questions in advance so that you can copy and paste the answers into the quiz.

1. Compare alternation of generations in plants with the animal life cycle. How does haploid/diploid dominance differ?

2. Which life cycle strategy (alternation of generations vs. animal life cycle) provides more opportunities for adaptation? Why?

3. Compare embryonic development in externally fertilized vs. internally fertilized organisms. What trade-offs are evident?

Figure Credits

Unless otherwise stated, figures are created and copyrighted by the authors or Chemeketa Press. © 2026.

Figure 3.1. "Fermentation tube" by creator BriannaBruno88. Quizlet, 2026.

Figure 5.1. "Model of intestinal villi," from Westrich & Berg, 2011. University of California Press.

Figure 5.2. "Determining the surface area of a glove finger" from Westrich & Berg, 2011. University of California Press.

Figure 5.3. "Diagram of the human digestive system (left), a cross-section of the intestine, and an up-close look at an intestinal villus (right)" from Cull, 1989. The Sourcebook of Medical Illustrations, public use.

Figure 5.4. "Generalized diagram of the structure of fish gills" from Cull, 1989. The Sourcebook of Medical Illustrations, public use.

Figure 6.5. "External and internal anatomy of a heart," from Neuroscience: Canada, 2nd Edition by Charles Molnar and Jane Gair. Open Library. Wikimedia Creative Commons Attribution 4.0.

Figure 7.2. "Ovarian cortex in a rhesus monkey," by author Tulemo. Wikimedia Creative Commons License 4.0.

Figure 7.5. "Lilium ovary L," by author Jon Houseman. Wikimedia Creative Commons License 4.0.

Figure 7.6. "Coleus stemtip," by author Jon Houseman. Wikimedia Creative Commons License 4.0.

www.ingramcontent.com/pod-product-compliance
Lightning Source LLC
Chambersburg PA
CBHW081300200326
41423CB00001BA/261